Cambridge County Geographies

General Editor: F. H. H. Guillemard, M.A., M.D.

SOUTH LANCASHIRE

Cambridge County Geographies

SOUTH LANCASHIRE

by

A. WILMORE, D.Sc., F.G.S., F.R.G.S.

With Maps, Diagrams and Illustrations

CAMBRIDGE

AT THE UNIVERSITY PRESS

1928

CAMBRIDGE UNIVERSITY PRESS
Cambridge, New York, Melbourne, Madrid, Cape Town,
Singapore, São Paulo, Delhi, Mexico City

Cambridge University Press
The Edinburgh Building, Cambridge CB2 8RU, UK

Published in the United States of America by Cambridge University Press, New York

www.cambridge.org
Information on this title: www.cambridge.org/9781107616165

First published 1928
First paperback edition 2013

A catalogue record for this publication is available from the British Library

ISBN 978-1-107-61616-5 Paperback

CONTENTS

PAGE

1. Lancashire a County Palatine: The Duchy of Lancaster 1

2. Position and General Characteristics . . 3

3. Size and Boundaries 5

4. Geology 6

5. The Build of South Lancashire . . . 19

6. Rivers 23

7. Coast-line 28

8. Climate 34

9. Natural History 38

10. People 51

11. History 57

12. Industries 69

13. Famous and Historic Buildings . . . 89

14. Communications 99

15. The Roll of Honour 111

16. The Chief Towns and Villages of South Lancashire 126

ILLUSTRATIONS

PAGE

Barley with Pendle in the background   .   .   .   12

Pendle from Clitheroe side   20

Sandhills at Ainsdale   31

Pier Head, Liverpool   33

The Heron   43

Cranesbill (*Geranium sanguineum*)   47

Prehistoric Implements   52

Roman Altar, found in Manchester in 1831   .   .   54

Whalley Abbey   62

Brungerley Bridge, Clitheroe   64

Clitheroe Castle   66

Spinning   82

Weaving   83

Dyeing in the Hank or Skein   84

Raising Cloth   87

Hoghton Tower   91

The Holme, Burnley   92

Hall-i'-th'-Wood Museum   92

Parish Church, Rochdale   94

Parish Church, Middleton   94

South Porch, Parish Church, Middleton .   .   .   96

Old Stocks, Colne Parish Churchyard   .   .   .   97

Town Hall, Bolton   98

PAGE

Royal Exchange, Manchester 100

Swing Aqueduct and Road Bridge, Barton . . 107

Steamship discharging in No. 9 dock, Manchester Docks 108

Sir Richard Arkwright 113

Samuel Crompton, inventor of the Mule Spinning Frame 114

John Dalton's birthplace 116

Thomas de Quincey 118

St Mary's Parish Church, Blackburn . . . 128

Market Cross, Bolton 129

Hall-i'-th'-Wood, Bolton 130

Municipal College, Burnley 132

Pack Horse Bridge, Wycoller 134

Wycoller Hall, near Colne 135

Castle Street, Liverpool 140

Liverpool Cathedral 141

The John Rylands Library 145

The Infirmary, Oldham 147

Lord Street, Southport 151

Technical College, Wigan 155

Diagrams 157

MAPS AND SECTIONS

Map showing the relation of South Lancashire to the
whole county . . . . . . PAGE 4

South Lancashire, showing the structure and boundaries   5

South Lancashire, a Geographer's Geological Map   .   7

A boring for coal showing the strata passed through .   11

Horizontal Geological section across the Clitheroe
Anticline, the Pendle Range and the Burnley
(or Calder) Basin   14

Horizontal Geological section across the Burnley (or
Calder) Basin and the Rossendale Anticline   .   14

The Calder and Darwen Rivers and the industrial
towns in their basins   24

The Irwell and its chief tributaries, with part of the
Ship Canal   27

Rainfall Map, England and Wales   36

The probable chief Roman Roads of South Lancashire
and district   58

Map showing the relation of the Chemical Industries
to raw material, fuel, transport, markets   .   .   74

Map showing the distribution of the Textile and allied
industries of South Lancashire   85

Map of the canals of South Lancashire and district   .   105

The former Lancashire and Yorkshire Railway—northern
part of South Lancashire   110

Railways of the southern part of South Lancashire not
part of the L.Y.R. system   111

The illustrations on pp. 12, 31, 33, 140, 151 are from photographs by Messrs Valentine and Son, Ld.; those on pp. 20, 64, 66 from photographs by Mr E. Buck; that on p. 43 from a photograph by Mr J. Holmes; that on p. 54 from a photograph by Mr J. J. Phelps; those on pp. 62, 91 from photographs by Mr A. Bellingham; that on p. 47 from a photograph by Mr A. Horner is reproduced by permission of Mr R. Farrer and Mr E. Arnold; those on pp. 82, 83 are from photographs by Mr A. W. Muncuster. The illustration on p. 84 was supplied by Messrs Burgess, Ledward and Co., Ld.; that on p. 87 by Messrs Samuel Heap and Sons, Ld.; the upper one on p. 92 is from a photograph by Mr J. Winskill; those on pp. 92 (lower), 98, 129 from photographs supplied by the Town Clerk, Bolton; those on pp. 94, 147 are from photographs by Messrs Allen and Sons, Blackpool. I am indebted to the Rector of Middleton and Mr H. Partington for the illustration on p. 96, and to Mr A. Rought Brooks for that on p. 97; that on p. 100 is reproduced by courtesy of the Master; and those on pp. 107 and 108 by courtesy of the Secretary, Manchester Ship Canal. That on p. 116 is from a photograph by Mr H. Bell; Mr H. Smith supplied that on p. 128: Mr A. R. Pickles, Director of Education, Burnley, that on p. 132; Mr R. T. Lawson that on p. 135. The illustration on p. 141 is from a photograph by Mr F. R. Yerbury; that of the Rylands Library was kindly supplied by the Librarian; and that on p. 155 is reproduced by courtesy of Mr G. Guest, Director of Education, Wigan.

1. Lancashire a County Palatine: The Duchy of Lancaster.

This book deals with the Geography of one of the most important and thickly peopled parts of Britain. Modern Britain is divided into counties, some of which are called *shires*. In official documents Lancashire is usually called "The County of Lancaster."

The name *County* means a region of which a count or an earl is the governor. The word *earl* is Anglo-Saxon; *count* is Norman-French. The word *shire* is Anglo-Saxon, and means a region *shorn* or cut from an Anglo-Saxon kingdom. England, before the coming of the Normans, was already divided into *shires*. The Norman conquerors made use of these shires and later treated them as counties. Lancashire is not mentioned in Domesday Book, because there was no such shire among the Anglo-Saxon divisions. Lancashire north of the Ribble was surveyed with Yorkshire; South Lancashire was mentioned as "Inter Ripam et Mersham" (between Ribble and Mersey) and was associated with Cheshire.

Manchester, Ribchester, and Lancaster were fortified stations or towns on the great Roman roads which ran to the north; the latter name meaning the fortified camp on the Lune. The Roman rule came to an end in the fifth century, and there followed that long period of disorder which we call "The Dark Ages." During this time the western slopes of the Pennines and the western plains were unsettled, and the history of the west is quite fragmentary. It is not until 1115 to 1118, that we meet with the "Honor of Lancaster"; and it is fifty years later, in 1168–9, that we find a mention of the County of Lancaster as contributing 100 marks to the Royal Exchequer. Lancashire, or the County of Lancaster, was thus a much later creation than many of the English counties. It does not seem to have been fully recognised as a county until about 1193–4.

In the middle of the thirteenth century Henry III made the County of Lancaster into an Earldom. A century later, in 1359 or 1360, Edward III raised it to the dignity of a duchy in favour of Henry, Earl of Lancaster. The first Duke of Lancaster had a daughter, Blanche Plantagenet, and she married her cousin, John of Gaunt, the fourth son of Edward III; these were both of Royal blood. In 1362 we find John of Gaunt spoken of as the *Duke* of Lancaster. During the years 1359–63, the county was made into a County Palatine, that is a county with Royal privileges.

The duchy was thus Royal, and in 1376 we read of a Chancellor whose duty it was to look after the revenues of the Royal Duke. In 1396 an Act of Parliament settled the privileges of the County Palatine on John of Gaunt and his successors for ever. Finally, in 1399, a son of John of Gaunt became King of England as Henry IV, and thus the revenues of the Royal duchy became merged in those of the Crown.

The revenues of the Duchy of Lancaster are still independent of Parliament, but an account has to be rendered annually. This is the duty of the Chancellor of the Duchy, an office which dates from the fourteenth century. In 1924 the revenues of the duchy, mainly from court fees, dues, and royalties, were £141,066, out of which, after paying expenses of management, £11,039, a salary of £2000 to the Chancellor, and making allowance for charities, etc., the sum of £70,000 was paid over to the Keeper of the Privy Purse of His Majesty the King.

2. Position and General Characteristics.

In the series of Cambridge County Geographies Lancashire is divided into two parts. This book deals with that part of the county which lies between the Ribble and the Mersey. Professor J. E. Marr has described Lancashire north of the Ribble in a separate volume of the series, under the title of *North Lancashire*.

South Lancashire is a roughly rectangular region bounded by the Irish Sea on the west, the Ribble on the north, an irregular county boundary on the east, and the Tame and the Mersey on the south. It includes the old "Hundreds" of Blackburn, Leyland, Derby, and Salford. A hundred consisted of a number of townships and later a number of parishes.

There are two distinct parts of South Lancashire— different in structure, climate, agriculture, and density of population. If a line be drawn from Preston to Warrington, to the east of it is the hilly region, where the rainfall is greater and the soil less fertile; to the west of it is the plain, where the rainfall is less and agriculture more successful. It may roughly be said that there is a manufacturing east—the land of the hills; and an agricultural west—the land of the plain.

Map showing the relation of that part of the county
treated in this work to the whole county

3. Size and Boundaries.

South Lancashire is about 48 miles long from east to west, that is from Blackstone Edge to Formby Point; its

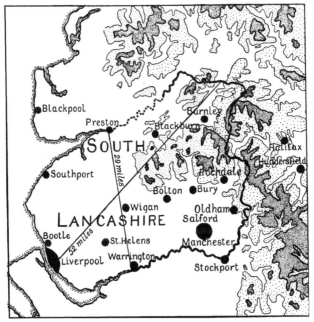

South Lancashire, showing the structure and boundaries.
The cities and county boroughs are shown

Land over 600 ft. lightly stippled; over 1200 ft. closely stippled

breadth from north to south, measured from Preston to Warrington, is about 29 miles. The longest straight line that can be drawn is from near Colne, in the north-east, to

Liverpool in the south-west, and measures about 52 miles. The area is about 1160 square miles; the population about four and a half millions.

At Mitton, near Whalley, two tributaries join the Ribble; these are the Hodder on the right bank, the Calder on the left. The Hodder does not concern us in this book. Going upstream from the confluence the Ribble is for several miles the county boundary, except at certain points where there has been some change in the course of the winding river. About three miles east of Clitheroe, near the ruins of Sawley Abbey, the county boundary leaves the Ribble and follows Ings Beck. It then follows the Pendle Range of hills for some distance, leaving this ridge at White Moor to cross one of the entries of the Craven Gap at Foulridge. It continues east of Colne and then runs almost due south for many miles along the Southern Pennines to a point about three miles east of Ashton-under-Lyne. From the confluence of the Ribble and Hodder at Mitton to the point mentioned just now the boundary line separates South Lancashire from the West Riding of Yorkshire. The boundary now follows the Tame to Stockport, and afterwards the Mersey to the Irish Sea. Along this latter part, from the point where it begins to follow the Tame, the county boundary separates South Lancashire from Cheshire.

4. Geology.

Before we discuss the build of South Lancashire it is advisable to learn something of its geology. Geology is the study of the rocks of the Earth's surface, how these rocks have been formed, the fossils they contain, and what the rocks and fossils tell us of the past history of the earth. By rocks we mean not only hard stone such as the grit-stone of Rossendale and of the Pennines but also the soft sandstones and sands of the Ormskirk district; even the loose sands of Formby and Southport are rocks to the geologist.

There are three chief classes of rocks accepted in geology.

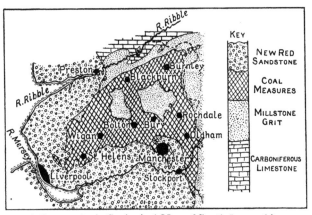

A Geographer's Geological Map of South Lancashire

(*a*) *Sedimentary Rocks.* These have usually been formed under water, layer upon layer, in estuaries, shallow seas, or lakes, and often contain the fossil remains of plants and animals which lived at the time when the rocks were laid down.

(*b*) *Igneous Rocks.* These have solidified from the molten condition, under varied circumstances and in very many forms.

(c) *Metamorphic Rocks.* These are very much altered rocks which have been made what they are by the long-continued action of heat, pressure, and chemical agents on other rocks.

South Lancashire has rocks of the first class only—the Sedimentary Rocks. No igneous or metamorphic rocks are met with anywhere except in the form of odd boulders or fragments which were scattered *on the surface* by the glaciers of the recent Ice Age. These scattered boulders will be mentioned again.

The **Sedimentary Rocks**, as already explained, were laid down in layers or beds one on the top of the other, and the recognition of this process and of the fossils contained is of the greatest assistance to the geologist, as he can thus learn something of the *relative* age of the rocks with which he has to deal. He knows very little about the *actual* age in years, but the *relative* age is of great importance to him. He divides up his sedimentary rocks into great groups or systems, each system consisting of the rocks deposited or laid down in a Geological Period. A number of systems make up a greater group still, and three such great groups are usually recognised in geology.

The oldest group of rocks is the *Primary* or **Palaeozoic**, a group containing six or seven systems, and the oldest of these systems contains the oldest fossil-bearing rocks we know. The *Secondary* or **Mesozoic** group contains three systems, and the *Tertiary* or **Cainozoic** three (or four), each representing much shorter periods in geological history than the **Palaeozoic** systems. Here is given a table of the groups and systems of rocks, mainly as we know them in Britain, with some examples of the actual rocks which are found in each system in the British Isles.

NAMES OF SYSTEMS		SUBDIVISIONS	CHARACTERS OF ROCKS
TERTIARY	Recent and Post-Pliocene	Metal Age Deposits Neolithic ,, Palaeolithic ,, Glacial ,,	Superficial Deposits
	Pliocene	Cromer Series Weybourne Crag Chillesford and Norwich Crags Red and Walton Crags Coralline Crag	Sands chiefly
	Miocene	Absent from Britain	
	Eocene	Fluviomarine Beds of Hampshire Bagshot Beds London Clay Oldhaven Beds, Woolwich and Reading Thanet Sands [Groups	Clays and Sands chiefly
SECONDARY	Cretaceous	Chalk Upper Greensand and Gault Lower Greensand Weald Clay Hastings Sands	Chalk at top Sandstones, Mud and Clays below
	Jurassic	Purbeck Beds Portland Beds Kimmeridge Clay Corallian Beds Oxford Clay and Kellaways Rock Cornbrash Forest Marble Great Oolite with Stonesfield Slate Inferior Oolite Lias—Upper, Middle, and Lower	Shales, Sandstones and Oolitic Limestones
	Triassic	Rhaetic Keuper Marls Keuper Sandstone Upper Bunter Sandstone Bunter Pebble Beds Lower Bunter Sandstone	Red Sandstones and Marls, Gypsum and Salt
PRIMARY	Permian	Magnesian Limestone and Sandstone Marl Slate Lower Permian Sandstone	Red Sandstones and Magnesian Limestone
	Carboniferous	Coal Measures Millstone Grit Mountain Limestone Basal Carboniferous Rocks	Sandstones, Shales and Coals at top Sandstones in middle Limestone and Shales below
	Devonian	Upper Mid } Devonian and Old Red Sand- Lower } stone	Red Sandstones, Shales, Slates and Lime- stones
	Silurian	Ludlow Beds Wenlock Beds Llandovery Beds	Sandstones, Shales and Thin Limestones
	Ordovician	Caradoc Beds Llandeilo Beds Arenig Beds	Shales, Slates, Sandstones and Thin Limestones
	Cambrian	Tremadoc Slates Lingula Flags Menevian Beds Harlech Grits and Llanberis Slates	Slates and Sandstones
	Pre-Cambrian	No definite classification yet made	Sandstones, Slates and Volcanic Rocks

The rocks of South Lancashire are confined to four of the systems in the table. The most important are the rocks of the great Carboniferous System. A glance at the geological map at the end of the book will show that these rocks occupy practically all the eastern part of South Lancashire. The Carboniferous System is sub-divided into formations which are of considerable thickness. Here they are given in tabular form, so far as they occur in South Lancashire and the Pennines generally.

Carboniferous System
(4) *Coal Measures* : a series of sandstones, shales, clays, and coal-seams.

(3) *Millstone Grit* : a series of coarse sandstones, flagstones, and shales.

(2) *Carboniferous* (or *Mountain*) *Limestone*, including "*Pendleside Series*" : a series of grey and dark limestones, and shales.
....
(1) *Basal Carboniferous Rocks* : not seen in South Lancashire.

The Carboniferous Limestone or Mountain Limestone is the oldest formation seen in South Lancashire. Its base is not apparent anywhere as there is no natural section or artificial cutting that has gone deep enough. These limestone rocks are seen only in the north-eastern corner of South Lancashire, in the district near Clitheroe. At Clitheroe and Chatburn there are several quarries from which limestone and lime are sent out in great quantities. The Carboniferous Limestone rocks are probably 5000 feet in thickness in that district. The lowest rocks are seen at Chatburn, and are

dark, muddy shales with dark-blue limestones. The limestone quarried at Chatburn is known as the Chatburn Limestone; it is much used for road-metal. The middle part

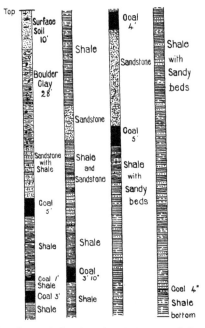

**A boring for coal showing the strata passed through,
and illustrating what is meant by coal measures**

Depth of whole boring 320 *ft.*

of the Carboniferous Limestone contains thick beds of
light-grey limestone, often crowded with fossils; this is
called the Clitheroe Limestone. It forms a fine series of
knolls or low hills about which there has been some keen

Barley with Pendle in the background

controversy among geologists. On one of these knolls the ruins of Clitheroe Castle stand. The upper part of the Carboniferous Limestone consists of muddy shales with beds of limestone of variable thickness and character. The most important of these limestones are grouped together as the Pendleside Limestone, which forms a distinct feature on the north-western face of Pendle. The beds on the northern face of Pendle, and beds of the same age in the mid-Pennines have been called the Pendleside Series by some geologists.

The *Millstone Grit* series is the middle formation of the great Carboniferous System in South Lancashire, and it is of great importance in the geology and geography of the county. It is at least 5000 feet in thickness in the neighbourhood of Pendle and Colne. It consists of a series of thick beds of coarse sandstone or grit with shales between. The lateral range of hills known as the Pendle Range is mainly built of coarse grits. The beds of the Millstone Grit formation dip or slope from Pendle under the Burnley Basin and come to the surface again in the hills on the southern side of the syncline in which Burnley and Nelson lie.

Between two of the massive grit beds in the Pendle district the Sabden Shales occur; these are nearly 2000 feet in thickness in North-east Lancashire, and give rise to a long line of "valleys" between the grit hills.

The name Millstone Grit comes from the early days of geology. Grit is a coarse sandstone composed mainly of hard, angular grains. In the days when nearly every Pennine valley had its small corn-mill driven by water-power

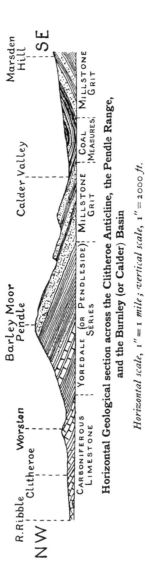

Horizontal Geological section across the Clitheroe Anticline, the Pendle Range, and the Burnley (or Calder) Basin

Horizontal scale, 1″ = 1 mile; vertical scale, 1″ = 2000 ft.

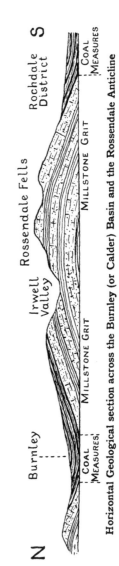

Horizontal Geological section across the Burnley (or Calder) Basin and the Rossendale Anticline

Compare with the above. This section is generalised and diagrammatic

the grit was used for making millstones; hence the name
Millstone Grit.

A geological formation often varies considerably in
character when it is traced from one district to another.
This is the case with the Millstone Grit. There are few
flagstones in the Pendle district where the formation is so
well developed, but in the Rossendale district there is a fine
series of hard, well-bedded flagstones called the Rossendale
Flags. These have been extensively quarried and have been
of great use in Lancashire. The uppermost thick bed of grit
is known as the "Farewell Rock" in Lancashire because
the miners learnt that they could bid farewell to any pro-
spect of finding important coal seams when they got down
to this grit bed. It is also known as the "Rough Rock," a
name given to it because of its mineral character.

The *Coal Measures* are the uppermost of the formations
of the Carboniferous System; they consist in Lancashire of
a thickness of about 4500 feet of sandstones, shales, fire-
clays, and coal seams. All the rocks are of economic importance but, of course, the coal seams are the most valuable.
A *coal seam* is a layer or bed of coal which usually maintains
a fairly regular thickness over a wide area. For example, the
Lower Mountain Mine, one of the well-known seams, is
usually between 3 feet 6 inches and 4 feet 3 inches in thickness. The coal measures are divided into Lower, Middle,
and Upper, the middle containing the greatest number of
good, workable seams of coal.

The sandstones of the coal measures often make good
building stones and many East Lancashire towns and villages are largely built of them; we may mention Colne,

Nelson, Burnley, Accrington, Haslingden, Rawtenstall, Bacup, Rochdale, and Oldham. The shales of the coal measures are much used for the manufacture of bricks, tiles, and sanitary ware; and the fire-clays for making fire-bricks and retorts.

A region where the coal measures occupy the surface is a *Coal Field*; South Lancashire contains a large and important coal field, which stretches from Colne in the north-east to the west of St Helens, a distance of over forty miles. In the middle of the coal field there are two patches in the Rossendale Fells where the coal measures are missing, and the Millstone Grit forms the surface rock. The coal measures have been worn away from these upland parts during the long geological ages since the Carboniferous period.

We must refer to the geological map again and also to the table of rocks on p. 9. From Manchester westward to Liverpool, and from the latter city to Preston we see that a belt of newer rocks occurs; they are nearly all Triassic, that is they belong to the oldest system of the Mesozoic group of rocks. The name Trias was given to this system because in Central Europe it consists of three formations; only two of these are represented in South Lancashire, known as the Bunter and the Keuper. The rocks are red or yellowish or mottled sandstones and marls, and the greater part of the plain of South Lancashire is underlain by them. Near Liverpool good building stones are obtained from the Trias; and in the Ormskirk district soft red sands are quarried for use as iron-moulders' sand.

There are also some patches of Permian sandstone in South Lancashire. Geographers frequently group the rocks of this system with those of the Trias under the name New Red Sandstone. The New Red Sandstone rocks usually yield open and fertile soils, and they underlie some of the best farming districts in the British Isles. Hence much of the west of our county division which is occupied by them is rich farming-land.

There are no rocks older than the Carboniferous in South Lancashire, and no Secondary or Tertiary rocks above the Trias. The newest system (see table) is represented by several kinds of superficial deposits. On the hill-sides and in the valleys of East Lancashire, and on many parts of the plain of West Lancashire, there are irregular deposits of a tough clay, bluish-black or yellowish in colour, and often full of scratched stones; this is called "boulder clay," and it was formed by the glaciers which came from the North and over-rode most of Lancashire during the Glacial Period. Scattered along the hill-sides and in many of the upland valleys there are patches or fragments of the old lateral and terminal moraines of the glaciers. In these moraines, and scattered over many parts of the surface—both on the hill-sides and in the valleys—there are found numerous fragments of rocks which do not belong to South Lancashire in the ordinary way. Among these are boulders of limestone from the Northern Pennines, bits of slaty rock and dark green grit from the neighbourhood of Ingleton, pieces of granite and other rocks from the Lake District, and fragments which have obviously come from the south-west of Scotland. Such rocks are called *erratics* because they have wandered far from their original home, and they are among the surest signs that a district was once glaciated. One kind of erratic is worth special mention. The lonely island of Ailsa Craig in the wide entrance of the Firth of Clyde contains an igneous rock in which are crystals of a comparatively rare, blue mineral called riebeckite. This rock can be recognised very readily, and it is interesting to note that it has frequently been found in South Lancashire.

Other recent deposits are the beds of peat on the eastern fells and in the peat-mosses of the plain. Peat consists of partially decayed remains of plants which grow best under swampy conditions. On the Lancashire fells it consists in the main of mosses, ferns, rushes, and grass, which by growth and decay, year after year, have gradually accumulated a considerable thickness of brownish-black peat.

The old meres of the Lancashire plain were probably the result of the Ice Age. The thick and irregular beds of boulder clay in the plain are impervious to water, and shallow pools or meres existed in many parts of the plain until quite recent times. They have all been more or less filled with the remains of aquatic vegetation which grew and decayed, the process probably continuing for thousands of years; this decomposed vegetable matter is the peat of the famous "peat-mosses." Chat Moss and Barton Moss are possibly the most famous, but there are many others. These peat-mosses occasionally contain the remains of trees, among which is the much-prized "bog-oak." Antlers of the Great Irish Elk are found in it from time to time, with other relics of the period which closely followed the disappearance of the ice.

Among the recently formed "rocks" we must include the muds and sands brought down to their estuaries by the Ribble and the Mersey, and to which a good deal of sea-drifted material is added. Immense deposits of these have been formed on the coast of South Lancashire, and much of the ground on which Southport stands has been thus built up during the last three hundred years.

5. The Build of South Lancashire.

We have already indicated that the build of South Lancashire is comparatively simple. The *eastern half* is occupied by the Pennines, the Pendle Range, and the Rossendale Fells. This is a land of broad moors and deep-cut valleys. The fells or moors are often bleak, wind-swept, and barren. The valleys were no doubt well-wooded before the trees were ruthlessly cut down to make room for houses, cotton-factories, and workshops.

These wide, breezy moorlands are of priceless value to the dwellers in the crowded industrial districts on their edges. Burnley, Blackburn, Bolton, Bury, Rochdale, and Oldham are all situated close to these moors, and these big, modern towns owe much to the pure air, the abundant water, and the opportunities for recreation provided by the wide fells.

The three divisions of the eastern moorlands have been indicated already. We may now look at them more closely. The first to be considered is that part of the Pennines which immediately concerns South Lancashire. A gap of lowland occurs in the middle of the Pennines through which it is possible to go from the western plain to the eastern plain by rising to very slightly over 500 feet. This well-known gap has been variously named the Skipton Gap, the Aire Gap, the Craven Gap. The name is not of very great moment, but, on the whole, the last is the best.

If we begin at the Craven Gap and go due southward we are on moorlands for about fifty miles; these are the Southern Pennines, some of the separate hill masses of

Pendle from Clitheroe side

which are partly in Yorkshire, partly in South Lancashire. The following are some of these border hills: Combe Hill, east of Colne, 1459 feet; Boulsworth Hill, south-east of Nelson, 1700 feet; Black Hameldon, south-east of Burnley, 1545 feet; Blackstone Edge, east of Littleborough, 1553 feet. These are all typical South Pennine moorlands—wide, bleak, and bare, with thick beds of peat on their upland expanses.

From the same low ground in Craven there runs from
north-east to south-west a well-defined ridge of hills. This
is the Pendle Range, which we can readily trace on the
map from Weets near Barnoldswick to Hoghton, south-
west of Blackburn. Weets, just in Yorkshire, rises to 1250
feet. Rimington Moor, across which the county boundary
runs, is 1253 feet high. Pendle, the highest point not only
in the Pendle Range but in South Lancashire, is 1831 feet,
and Spence Moor 1498 feet above sea-level. This line of
hills overlooks the Ribble valley on the one side and the
valley of the Lancashire Calder on the other.

The Rossendale Fells stand out like a western bastion
from the Pennines. There is no sharp dividing line between
the two, but we may conveniently take the valley of the
two Calders from Burnley to Todmorden for that purpose,
and include all the moors west of that line in the Rossendale Fells. As thus defined they form a roughly rectangular
mass of broad moors, cut up by numerous deep valleys.
The following are some of the best known "hills" in this
group of moorlands: Hameldon, south-west of Burnley,
1305 feet; Great Hameldon, east of Accrington, 1343 feet;
Rough Hill, north of Rochdale, 1419 feet; Hailstorm Hill,
north-west of Rochdale, 1525 feet; Bull Hill, north of
Ramsbottom, 1372 feet; Cartridge Hill, south-west of
Darwen, 1317 feet; Winter Hill, south-west of Bolton,
1498 feet; and Rivington Pike, north of Horwich,
1190 feet.

Among these East Lancashire Fells are many narrow
valleys, frequently of considerable beauty. From nearly
every large town on the edge of the Fells it is possible to

make short excursions into woods sheltered in the narrow "cloughs" or "denes." These bits of woodland are much frequented by the factory workers of the industrial towns and picnics are a common feature in the summer.

The wide tops of the moors are usually underlain by either Millstone Grit or sandstones of the coal measures, with a covering of coarse grass and heather and much peat, as already mentioned. Here and there is a dwarfed mountain ash, or a willow, or a hawthorn bush, struggling against adverse circumstances.

To the west of these hills is the South Lancashire Plain, which is underlain for the most part by the sandstones and marls of the New Red Sandstone, though some parts are underlain by the coal measures. This plain forms a curved belt round the Rossendale Fells and the south-western end of the Pendle Range. A traveller by rail may see the character of it very well on a journey from Manchester Exchange Station to Liverpool, using George Stephenson's original railway route across Chat Moss; and then by travelling north from Exchange Station, Liverpool, through Ormskirk to Bamber Bridge.

In earlier times a good deal of the South Lancashire Plain was impassable moss, but most of the "mosses" have now been drained and converted into profitable, agricultural land. The drier parts of the plain were originally covered with forest, and Domesday Book tells us that there were 250,000 acres of dense woods between the Mersey and the Ribble in 1086; this is about one-third of the whole area. Extensive woods still remain in the "parks" of some of the county seats scattered over the plain.

The western part of the plain reaches the sea in a coast absolutely without rocky cliffs; and from Preston round by Formby Point to the Mersey entrance at Widnes is low, flat coast, made up of sands or clays.

6. Rivers.

With two trifling exceptions the whole of South Lancashire is drained to the Irish Sea. In the north-eastern corner there is a square mile or so, including the village of Foulridge, which drains to the Aire and so to the North Sea. South-east of Holme, near Burnley, the source of the Yorkshire Calder is in Lancashire. Two rivers and their tributaries carry the drainage of South Lancashire to the sea: the Ribble and the Mersey. The River Douglas (or Asland) may be treated as a tributary of the Ribble.

The Ribble rises on one of the plateaux of the Northern Pennines, in the high ground between Penygent and Ingleborough; two moorland streams, Gayle Beck and Cam Beck, uniting to form it. The river flows past Horton-in-Ribblesdale to Settle, where it leaves the Craven Highlands. It then crosses the Craven Lowlands, past Long Preston, Hellifield, Gisburn, and Sawley Abbey, from which to its junction with the Hodder at Mitton, near Whalley, it forms the county boundary. After Sawley Abbey has been passed all the left-bank tributaries are connected with South Lancashire. The first of these is Ings Beck, a pretty stream, which also forms the county boundary. A number of becks drain the northern face of Pendle, and run through the series of villages, Downham, Chatburn, Worston, Pendleton, and Barrow, and enter the Ribble as left-bank feeders.

About two miles north-west of Whalley the Lancashire Calder joins the Ribble. This river drains the district of Accrington, Burnley, Nelson, and Colne. Some distance east of Colne rises the river Laneshaw; which at Laneshaw

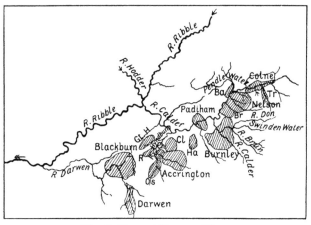

**The Calder and Darwen Rivers
and the industrial towns in their basins**

Ba = Barrowford
Br = Brierfield
Ch = Church
Cl = Clayton-le-Moors
Gt H = Great Harwood

Ha = Hapton
Os = Oswaldtwistle
R = Rishton
Tr = Trawden

Bridge is joined by Wycoller Brook and is now called Colne Water. It is joined later by Trawden Brook and other small streams, and after passing through Colne unites with Pendle Water. The latter stream has come from the southern slopes of Pendle, having flowed through pretty scenery past Barley, Roughlee, and Barrowford. The united

stream flows past Nelson and Brierfield and joins the stream
which bears the name of the Calder in the low ground
north of Burnley. The Calder rises at Holme, south-east
of Burnley, near the source of the Yorkshire Calder; it
flows past historic Towneley Hall and through the middle
of Burnley, where it is joined by the Brun. This stream
has come from bleak Pennine fells past Hurstwood, and
has united with Swinden Water and the Don. The latter
runs from the wild, romantic valley of Thursden near the
county boundary.

The Calder flows on through Padiham, and a little later
the drainage of Accrington is brought to it by means of
the River Hyndburn. Below the confluence with the Hyndburn the Calder cuts its way through the Pendle Range at
the Whalley Gap. Just through the gap, on a flat meadow
by the Calder, are the ruins of the Cistercian Abbey of
Whalley.

The River Darwen is the next important left-bank tri-
butary of the Ribble. It collects the drainage of the north-
western edge of the Rossendale Fells, and flows through
Darwen and Blackburn. At Hoghton it cuts through the
Pendle Range in the same way that the Calder does at
Whalley and a picturesque gorge is the result.

The Ribble opens out into a wide estuary below Preston,
but before the open sea is reached the Douglas River joins
it. The Douglas drains the western edge of the Rossendale
Fells. The main stream flows through Wigan and on to
Rufford, after which the united waters of the Yarrow and
the Ettrick join it, bringing the drainage of Adlington,
Chorley, Leyland, and Lostock Hall. The Douglas makes

a subsidiary estuary of its own where it joins the Ribble at Hesketh Bank.

The other rivers of South Lancashire are the Tame, the Mersey, and the Irwell. The Tame rises on the peat-covered moors south-east of Ashton-under-Lyne and flows westward in a winding course to Stockport; here it is joined by the Goyt, and the Mersey is formed. This river flows past Stretford and Warrington to its great and important estuary.

The Irwell and its feeders drain the greater part of the Rossendale Fells southwards. The Irwell itself rises not far north of Bacup, near a village with the appropriate local name of Irwell Springs. It flows through Bacup, Waterfoot and Rawtenstall in a narrow valley. Soon after passing Rawtenstall it receives the drainage of the Haslingden district, and proceeds through a wider valley past Rams-bottom to Bury. Beyond this town it enters the plain where, after winding about a good deal, it receives the Roche from the Rochdale district, and, passing Radcliffe, is joined by the drainage from the district round Bolton, brought to it by the River Tonge and Bradshaw Brook. The Irwell now reaches Salford and Manchester and for a time forms the artificial boundary between the two. At a point close by the Cathedral and Grammar School it is joined by the Irk which has come from Royton and Mid-dleton; and on the other side of Manchester, the Medlock comes in, bringing the drainage of South Oldham, Fails-worth, and Droylsden. The Irwell now flows across Barton Moss and unites with the Mersey a little west of Stretford.

The Irwell and its numerous tributaries are of great importance in the industrial life of Lancashire. The number of factories of various kinds on their banks is enormous, and the water of the river as it passes through Manchester

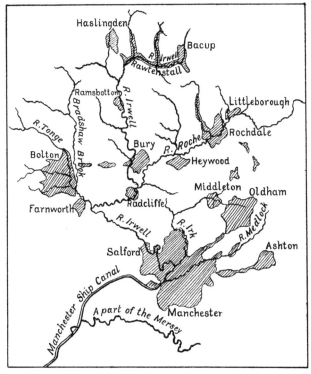

Map of the Irwell and its chief tributaries, with part of the Ship Canal

The more densely peopled areas are shaded

and Salford bears witness to the number of works which have poured their more or less filtered effluents into the streams.

Some distance west of Warrington the Mersey begins to widen out into its great bottle-shaped estuary. Here is a great basin twelve miles long and from two to three miles wide. Farther towards the sea—opposite Liverpool—the estuary narrows to three-quarters of a mile. The great basin is filled with water at the high or spring tides, and is half filled at the low or neap tides. The scour of the tides emptying from the basin helps to keep open the Liverpool channel, not only in the narrow "bottle-neck" but for some eight miles or so out to sea. The result is that there is a channel more than 50 feet deep opposite the Liverpool landing stage, and here the largest vessels can ride easily at any time. At the north end of the landing stage, even close to shore, the depth is 30 feet at low tide. Here the big liners come to anchor.

It may be noted that dredging is necessary to keep the port of Liverpool fully open for the largest liners, and the Mersey Docks and Harbours Board takes out 13,000,000 tons of material annually.

7. Coast-line.

The coast of South Lancashire is quite unlike that of many counties, as it can show no hard rocks, and nothing that really deserves the name of a cliff. It must not be imagined, however, that it is featureless; on the contrary, the record of recent coastal changes makes a story of profound interest.

We may begin at the entrance to the Preston docks, where the tidal Ribble begins to widen somewhat gradually.

The coast is here low and flat, with a good deal of fine river silt or alluvium. After about four miles the river widens more rapidly, and the River Douglas comes in on the south side from the South Lancashire plain. Here there have been extensive changes within historic times, and even within the last century. The present opening of the Douglas into the Ribble is certainly more than a mile farther north than it was in 1824. In the eighteenth century there was on the Douglas a port of some importance called Mylthorp, which was about two miles from the present river mouth. A new channel for the river was dug by the Ribble Navigation Company in 1853; this is used as a branch of the Leeds and Liverpool Canal, which joins the river at Tarleton.

Between the estuary of the Douglas and Southport begins that extensive belt of sand dunes or ridges of blown sand which continues to the northern suburbs of Liverpool and reaches its greatest width opposite Formby Point. Behind these sand dunes are many great stretches of peat, one of the best known and most extensive of which is Tarleton Moss. Here, up to the seventeenth century, existed a shallow fresh-water lake called Marton Mere[1] which is shown on old maps, such as that of Speed of 1610. Leland, the famous antiquary, described it in about 1544 as being about four miles long and three miles wide. The drainage of this lake was undertaken seriously in 1692, was continued by successive generations of landowners, and it is now wholly converted into agricultural land.

[1] There is a Marton Moss, the site of an old mere, between Kirkham and Lytham, in North Lancashire.

It is a common saying among Lancashire folk that the sea is leaving Southport; this is only partly true as will be seen. On the north side of Southport there have been large gains of land at the expense of the sea within the last few centuries. This new land consists largely of sand of the delta and river mud. The two rivers, Ribble and Mersey, drain inland areas where the coarse sandstones of the Millstone Grit and the fine sandstones of the coal measures are exposed; and immense quantities are carried down to their estuaries. The scour of the tidal currents on the South Lancashire coast is from south to north and much of the Mersey sand is swept along past the Formby angle to mingle with that of the Ribble. It is mainly this river-borne and tide-swept material that has given the new land to the north of Southport. From old maps it is learnt that where Lord Street now runs it was possible—as the Southport people are fond of saying—to catch shrimps 200 years ago. All the land between the street and the northern extension of the promenade seems to be a growth of the last two centuries.

Near the northern end of Southport there is a slight rise in the sandy ground, which was formerly known as Sugar Hillock. Tradition says that a ship with a cargo of sugar and potatoes was wrecked here in the early seventeenth century. Some of the tubers were planted in the district, and Churchtown (Southport) claims to be the first place, or at least one of the earliest, where potatoes were grown in England. It is perhaps fitting that the hinterland of this coast is now one of the most famous potato-growing regions in the British Isles.

South of Southport the line of the old coast comes closer
to the present one, and there is evidence that within the
last half-century the sea has here gained slightly on the
land. The great changes which have certainly taken place
within the last five hundred or six hundred years are due
to the interaction of three processes—the subsidence of the

Sandhills at Ainsdale

land, the erosive work of the waves, and the piling up of
river-borne and sea-drifted sand by the winds.

Two small towns, a number of hamlets, and many farms
have disappeared, as may be seen by reference to old records.
The most famous of the lost towns was Argarmeols, which
was probably situated somewhere in the neighbourhood of
the present Southport. It was lost in the early part of the
fourteenth century—which was a time when extensive

coastal changes took place in many parts of western Europe. A deed of Cockersand Abbey, dated 1346, refers to the place thus:— "Argarmeles, which is now annihilate by the sea and there is no habitation there."

Southward to Formby the sand dunes grow wider until opposite the blunt point or cape they are nearly three miles in width. It seems almost incredible that three centuries ago there were no sandhills at Formby, and yet we are assured that even so recently as 1690 this was the case. The name of the present village, Formby-by-the-sea, seems almost a joke to-day. Old Formby was an important port and fishing station; but little or nothing can now be seen of the old town. The channel of the port began to silt up about 1700, and on the new land thus formed dunes of blown sand accumulated with great rapidity.

South of Formby the river Alt enters the sea, making a curious bend to the south before doing so. At the mouth of this river there was a port, Altmouth, but this too has disappeared. The river has changed its course, and flows into the sea probably more than a mile to the south of its medieval estuary. Its course is changing rapidly at the present time. The blown sand practically ends at the North Wall Lighthouse where the "bottle-necked" estuary of the Mersey may be said to end seaward. The estuary opposite the Liverpool Landing Stage is only 1250 yards wide.

From the termination of the sand dunes there is a continuous line of docks for many miles, and beyond Liverpool to the south-east there are docks at Garston. The Mersey here has opened out into its wide pool. At the old-world village of Hale, and still farther from the open sea at the very modern, busy town of Widnes, the estuary may be said to end.

The Mersey estuary, with its magnificent ports of Liverpool and Birkenhead, is one of the most important river-mouths in Britain. Yet it is possible that it is of quite recent origin. Some authorities suggest that until the early

Pier Head, Liverpool

Middle Ages the Mersey flowed across the southern end of what is now the Wirral Peninsula of Cheshire to join the Dee, but this is not accepted by all. It is, however, of interest to note that the Romans never refer to the Mersey, and that Ptolemy, the great Roman geographer and map-maker, shows only one estuary between the Dee and Morecambe Bay. This he calls Belisama, a name which almost certainly belonged to the Ribble.

8. Climate.

Climate is average weather. When we have learnt something about the average direction of our winds, the average amount of rain, and the average temperature, we say we have studied our *climate*. All these factors of weather vary considerably from season to season and from year to year, but when we get an average for a prolonged series of years, we can make general statements of value. The three factors, winds, rainfall, and temperature, are, of course, interdependent, but it is convenient to take them in the order named.

Winds. Though our winds are exceedingly variable, everyone must have noticed that more winds blow from some westerly point than from any other. The observatories at Bidston, Manchester, and Stonyhurst all corroborate this everyday observation. This is a most important conclusion, because these commonest winds blow from over the North Atlantic; they therefore come laden with moisture which they are ready to drop on the Lancashire hills. Further, though there are more west winds on the whole than any other, there is a tendency towards an increase of south-west winds in winter and north-west winds in summer. The changes will clearly tend to raise the winter temperature and lower that of summer, thus making it more equable.

Rainfall. The warm moist winds, which come from the Atlantic, reach the South Lancashire plain first, and the hilly regions of East Lancashire later. The moving air

must rise to pass over these hills, and in so doing it becomes
cooled. The cooler air cannot contain as much moisture,
which therefore falls as rain on the Rossendale Fells and
the Western Pennines. We thus see why East Lancashire
is wetter than West Lancashire. Most of the western
plain has an average rainfall of 30 to 35 inches; the hilly
regions of the east have an average rainfall of 40 inches
or more, as will be seen from the general rainfall map of
England and Wales.

Much rain means much cloud, and East Lancashire
people realise how often their sky is covered with clouds
of some kind; there is more sunshine in the west of the
county and that is one reason why Southport and Formby,
in South Lancashire, are such favourite holiday resorts
among the workers of the densely peopled industrial districts
of East Lancashire.

The next point to be considered is how the rain comes
in the different seasons of the year. It is convenient to
take the seasons as in the following scheme:

Winter: December, January, February.
Spring: March, April, May.
Summer: June, July, August.
Autumn: September, October, November.

So far as figures are available the seasonal rainfall is
approximately as follows: Winter 28 $°/_o$, Spring 20 $°/_o$,
Summer 25 $°/_o$, Autumn 27 $°/_o$. From this it will be seen
that though there is no very great difference between the
seasons, spring is the driest.

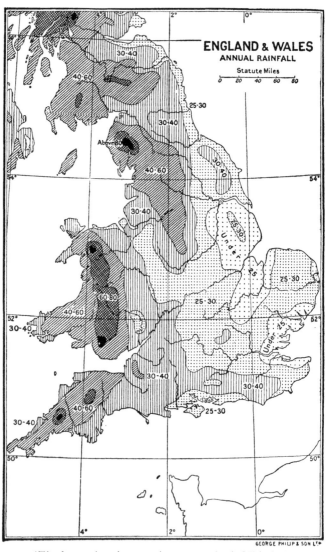

ENGLAND & WALES
ANNUAL RAINFALL

Statute Miles

0   20   40   60   80

30-40

40-60

25-30

30-40

Above 80

40-60

30-40

30-40

Under 25

30-40

25-30

60-80

40-60

25-30

30-40

Under 45

30-40

40-60

30-40

30-40

40-60

25-30

30-40

25-30

GEORGE PHILIP & SON LT.

(The figures give the approximate annual raiufall in inches)

Temperature. This depends very closely upon the direction of the prevalent winds and upon the rainfall. It is a common experience in East Lancashire in winter to have cold weather with an east wind, and then for a rapid rise of temperature to follow a change in the wind. In winter the east and south-east winds blow from a cold continent, the west and south-west winds from a warm ocean. The latter winds bring the warm rains also.

The average temperature of Great Britain and Ireland taken as a whole is about 50° F.; this is almost exactly the average temperature of Liverpool or Southport. The average January temperature of Great Britain and Ireland is about 40° F., and that of July 60° F., and again these are almost exactly the temperatures of West Lancashire. No part of the world *in the same latitude* has so high an average temperature as the British Isles. South Lancashire lies between 53° and 54° North latitude; here is a comparison with the average temperatures of some other places in the same latitude:

		January	July	Range
South Lancashire Plain	...	40° F.	60° F.	20° F.
Danzig (Baltic coast)	...	30° F.	64° F.	34° F.
Kamschatka coast	...	5° F.	52° F.	47° F.
Rigolet (Labrador)	...	7° F.	50° F.	43° F.
Fort Simpson (B. Columbia)		33° F.	58° F.	25° F.

It is obvious that the high average temperature and the low range are both of great importance in an industrial community such as South Lancashire.

So far the figures for the plain only have been taken, so that no complications should follow from the elevation. Much of the eastern half is hilly, however, and here it is cooler on the whole than the western plain, both in summer and in winter. It may be said, roughly, that in ascending from sea-level the air becomes 1° F. cooler for about every 300 ft. of ascent. The average elevation of Bolton and Bury is about 300 feet; we may therefore expect the average temperature of these places to be about 1° F. lower than that of Liverpool. The average elevation of Blackburn, Rochdale, Burnley, and Nelson is about 450 feet; the temperature should thus be about 1½° F. lower than that of Liverpool. The average elevation of Oldham and Colne is about 600 feet, which would make them about 2° F. cooler than the coast; and these figures accord well with actual experience.

On the Rossendale Fells, the Pendle Range, and the Pennines, snow is quite commonly seen when there is none in the plain. The moorland roads between Todmorden and Bacup, Blackburn and Bolton, Nelson and Clitheroe are often blocked by heavy snowfalls in winter. On the other hand, no great amount of snow falls on the western plain of South Lancashire, and it seldom remains very long there.

9. Natural History.

ZOOLOGY.

The prehistoric animals formerly inhabiting our county include the great Irish elk (*Cervus giganteus*), the urus (*Bos primigenius*), the mammoth (*Elephas primigenius*) and the cave bear (*Ursus spelaeus*). Magnificent antlers of the Irish elk have been found at places so far apart as Altham in the east and the estuary of the Ribble in the west. The wild boar, the wolf, and the red deer have become extinct within comparatively recent times. Until the seventeenth century the forests of Pendle, Trawden, and Rossendale were true forests, well suited to such animals. Wild boars were hunted in the Pendle district as late as 1617; though the wolf seems to have disappeared finally about the end of the sixteenth century. Herds of red deer persisted until well into the eighteenth century, and in the neighbouring region of Bowland they were in existence until about 1805.

The wild white cattle of Middleton Hall, Hoghton Tower, and Lords' Park near Whalley, seem to have become extinct by the end of the eighteenth century; at Gisburn, about four miles from the county boundary, the last survivor of a herd (of which a specimen is preserved in the Manchester Museum) was killed in 1857. The beaver (*Castor fiber*) is, of course, quite extinct, but that there must have been many communities of these animals existent within fairly recent times is shown by the name "Beaver" being used for so many upland farms and pastures.

The badger (*Meles taxus*) is said no longer to be found within our boundaries; at least there does not seem to be any authentic record of one within recent years. It was common enough three hundred years ago, as we learn from the interesting diary of Nicholas Assheton of Downham, written in 1617, 1618. The fox is now becoming rare in

South Lancashire. The otter is still found in the Ribble, occasionally in the Calder, the Darwen, and the Douglas, and perhaps more rarely in the river Alt.

Among the lesser carnivora are the weasel (*Mustela vulgaris*) which is abundant; the stoat or ermine (*Mustela erminea*) which is probably not quite so common, but frequently seen, and killed by gamekeepers; and the polecat (*Mustela foetida*). The latter is of course not numerous like the weasel, but it is said by good naturalists to be abundant in some limited localities.

The rodents are well represented in South Lancashire. The squirrel (*Sciurus vulgaris*) is common in the well-wooded districts of the New Red Sandstone plain, and by no means rare in the woods of the north-east of the county division. The dormouse (*Muscardinus avellanarius*) occurs in colonies in the western part of the county, but is seldom seen in the eastern part. The brown rat (*Mus decumanus*) is found everywhere and has become a great pest. This larger and more ferocious rat came into England from the East about 1720–1730, and has almost displaced the black rat (*Mus rattus*) which is now very rare indeed in South Lancashire. Three species of true mice are found: the house mouse (*Mus musculus*), the wood mouse or long-tailed field mouse (*Mus sylvaticus*), and the little harvest mouse (*Mus minutus*) which now seems to be very rare. Other small animals of the order Rodentia are the voles, of which there are three in our county division: the water vole or water rat (*Microtus amphibius*) is frequently seen in ditches, ponds and small streams; the field vole (*Microtus agrestis*) is fairly common and widely distributed; and the bank vole or red field vole (*Arvicola glareolus*) is abundant in the agricultural plain, where it spends its winter among the heaps of turnips and potatoes.

The hare, though still quite common, is thought to be diminishing in numbers; the rabbit is exceedingly abundant, especially among the sand dunes near the coast.

The Insectivora are represented in South Lancashire by five species. The mole is both abundant and widely distributed. It is known among the country people by several other names, among which Mowdywark and Mouldwarp are perhaps the commonest. The hedgehog is also abundant and widespread; it is, however, such a lover of the twilight and the darkness that it is not very often seen. The common shrew (*Sorex araneus*) and the water shrew (*Crossopus fodiens*), small, mouse-like animals which live on insects, snails, and worms, are quite common. The little pigmy shrew (*Sorex minutus*), which is comparatively rare, is the smallest of British mammals, having a body scarcely two inches in length.

Three species of bats are not at all uncommon in South Lancashire. These are the long-eared bat (*Plecotus auritus*), the pipistrelle (*Scotophilus pipistrellus*), and Natterer's bat (*Vespertilio nattereri*). The bats come out at dusk in summer, or on warm winter afternoons, and may then be seen near trees, houses, and barns flying in search of their prey, which consists of moths and other flying insects. The long-eared bat is probably the commonest, followed closely by the little pipistrelle. Four other species of bats have been recorded but they are all comparatively rare.

It has been said that 269 species of birds are known within the whole of Lancashire, of which 136 make their nests here. There are a few of these species which are rarely met with in South Lancashire, as they are birds of the Lakeland fells and crags of North Lancashire. It is obvious that only a few of this great number of species can be mentioned here.

As there are no rocks along the shore the common sea-birds do not nest so frequently as in some counties. The commonest species found on or near the coast are the black-headed gull (*Larus ridibundus*) and the lesser black-backed gull (*Larus fuscus*). The great crested grebe (*Podicipes cristatus*), and the little grebe or dabchick (*Podicipes fluviatilis*) are allied birds which nest regularly near the meres and reservoirs of the county.

Of the birds of prey we may mention the sparrow-hawk (*Accipiter nisus*) which is a resident and is still fairly abundant although every gamekeeper makes relentless war on it, the buzzard (*Buteo vulgaris*) which nests in the eastern uplands, the kestrel or windhover (*Falco tinnunculus*) which is also a common resident and is probably the best known of the hawks of South Lancashire. The merlin (*Falco aesalon*) is almost confined to the eastern moorlands, where bird-lovers frequently report its presence.

Three species of woodpecker are known in the county, the commonest being the green woodpecker or heyhough (*Gecinus viridis*) which is frequently found in the more thickly-wooded districts.

The kingfisher (*Alcedo ispida*) seems to be on the increase and is frequently seen quite close to villages and even large towns, where the streams provide it with its fish food.

The Heron

The heron (*Ardea cinerea*) frequents the canal reservoirs at Colne, Rishton, and Hollingworth, and is often seen wading in the meres, ditches, and rivers of the plain.

Some interesting birds of the moorlands are the red grouse (*Lagopus scoticus*), a species not found out of Britain, which is rigidly protected on some of the eastern fells; the snipe or heather bleat (*Gallinago coelestis*), which is abundant both on the fells and on the marshes of the west; the peewit or lapwing, also called tewit and green plover (*Vanellus vulgaris*), which is very common on the fells of the Pendle Range and Rossendale; and the curlew or whaup (*Numenius arquatus*), whose cry is rarely heard except in the more lonely parts of the moorlands. The dunlin or sea-lark (*Tringa alpina*) is a winter visitor which frequents the sandbanks of the west in countless thousands, but rarely nests in the county.

The Reptiles of South Lancashire.

There are five species of reptiles well known to naturalists, but no doubt many residents of South Lancashire have never seen one of them. The commonest reptile is the lively little sand lizard (*Lacerta agilis*) which may frequently be seen on the sandhills of the west. It is about 7 to 8 inches in length, the male dark greenish in colour, the female brownish-grey. The eggs are deposited in the sand to be hatched out by the heat of the sun. The viviparous lizard or swift (*Lacerta vivipara*) is much rarer but has probably a wider distribution, being found on the drier parts of the moors and mosses as well as on the sandhills.

The slow-worm or blind-worm (*Anguis fragilis*) is generally spoken of as a limbless lizard. It is about 9 to 10 inches in length, of a brownish grey colour with a medium black streak. It looks at its best when gliding silently through the grass on the hedge bank. The popular description is not strictly true, for it has small eyes, partly hidden by scales, and possesses rudimentary limbs hidden beneath the skin. It is not at all common in South Lancashire, but is still found in the plain and in the north-eastern part of the county.

Two snakes are found. The common or ringed snake (*Tropidonotus natrix*) is seen occasionally in West Lancashire and in the Whalley and Pendle districts. The viper or adder (*Vipera berus*) is very rare now, but is occasionally reported from the fertile Ribble valley near the Yorkshire border. This is our only poisonous reptile.

The Amphibia of South Lancashire.

There are five species of Amphibia found in the county, and four of them are abundant. The frog (*Rana temporaria*) is plentiful everywhere. The common toad (*Bufo vulgaris*) is also found all over the county. The natterjack toad (*Bufo calamita*) is not nearly so common, but may still be seen in the "slacks" among the sand dunes, and in the ditches and hedge bottoms of the interior.

Two species of newt are widespread. The common newt or smooth newt (*Molge vulgaris*) is also known by country people as the eft and the askard. It is found wherever there are ponds or meres in which it can lay its eggs. The great crested newt (*Molge cristatus*) is much rarer, but occasionally

in the ponds there may be seen fine specimens eight or
nine inches in length, with a rough warty back, and if
it is the breeding season having a fine crest and a bright
orange-coloured belly. This newt is a favourite object in
aquaria.

BOTANY.

The plants which grow in a district depend upon the
climate and the soil, and from this point of view we may
recognise five broad divisions in South Lancashire.

(i) The Limestone region in the north-east. Here the rocks
are mainly limestones and calcareous shales, with a variable
covering of boulder-clay The soil is fertile and plant life
varied and abundant. Among the more well-known characteristic wild flowers are the primrose, bird's eye primrose
(*Primula farinosa*), cowslip (*Primula veris*), meadow cranesbill (*Geranium pratense*), great bindweed or convolvulus
(*Convolvulus arvensis*), foxglove (*Digitalis purpurea*), honey-
suckle (*Lonicera Periclymenum*), bluebell (*Scilla nutans*),
garlic (*Allium oleraceum*), and arum or cuckoo-pint (*Arum
maculatum*), also known as lords and ladies. The great
Canterbury bell, which is supposed to have been introduced into England in the sixteenth century, is a beautiful
flowering plant found quite commonly in the hedgerows
about Downham. The bird's eye primrose loves the
marshy ground where boulder-clay overlies the Carboniferous Limestone. The meadow cranesbill is quite common
in this corner of Lancashire, but it has almost disappeared
from the neighbourhood of the big towns in the industrial
region. There are, of course, many other flowering plants

in the fields, hedgerows, and meadows of this limestone region, and the botanist is sure to find much to interest him in the neighbourhood of Twiston, Downham, Worston, Pendleton and Wiswell.

Cranesbill
(*Geranium sanguineum*)

(ii) The second type of country is that of the Millstone Grit and coal measures. Here the rocks are sandstones, grits and shales, which do not form soils so fertile as the rocks of the limestone region. Boulder-clay and patches of glacial sands and gravels, together with river gravels and finer silt, increase the fertility of districts which would otherwise be considered rather barren. The more interesting of these are Roughlee, Sabden, Wycoller, Towneley,

Whalley, Pleasington, and Rivington. Among the noteworthy plants of these districts are the orchids, the grass
of Parnassus (*Parnassia palustris*), the butterwort (*Pingui-
cula vulgaris*) and a species of sundew (*Drosera rotundifolia*).
Orchids, of several species, occur commonly on the clayey
and shaly banks of the mountain streams where there
is generally plenty of moisture. The beautiful grass of
Parnassus is generally regarded as one of the relics of
the Alpine flora left to us by the Ice Age. Butterwort
is very common in many of the upland valleys of the
Pennines and the Rossendale Fells: the sundew is much
rarer.

On the great moorlands are wide expanses of swampy
plateaux, the home of mosses, rushes, and cotton-grass.
The down-covered seeds of the latter plant (*Eriophorum
polystachyon*) make an interesting sight on many an East
Lancashire moor in late summer. On these same moors
sphagnum moss grows profusely. Another characteristic
form of vegetation is the bracken (*Pteris aquilina*) which
sometimes grows in immense profusion on the slopes of
the moors. On the drier parts there is a good deal of ling
(*Calluna vulgaris*); the bell heather (*Erica*) is much less
common. The bilberry or blueberry (*Vaccinium Myrtillus*)
with its myrtle-like leaves and bluish-black berries is also
found on the drier moors.

(iii) The third region is that of the New Red Sandstone plain of the west, which, like the eastern moors, is
plentifully scattered with boulder-clay and other glacial
deposits. Here are the most extensive woodlands in South
Lancashire, where we find the oak, beech, elm, ash, sycamore

and horse-chestnut. It is interesting to notice the difference between the oaks of the lowlands and the dwarfed, wind-swept specimens which are common in some of the upland valleys of the eastern part of the county. Under the shelter of the woods, there is, in spring, a profusion of bluebells and wood-anemones (*Anemone nemorosa*); in the more open parts such plants as the greater willow-herb (*Epilobium angustifolium*) are plentiful. There are, of course, many other interesting flowering plants, and most of those already mentioned from the other regions grow profusely in the plain.

A considerable area of the plain has once been covered by "mosses," as was pointed out in an earlier section, but many of these have now been drained. Barton Moss, Chat Moss, and Tarleton Moss still show us something of the vegetation which once accumulated through a long series of years. These mosses abound in ditches which act as drains and in these many water plants may be studied, among which are the bladder-wort (*Utricularia vulgaris*), the frogbit (*Hydrocharis morsus-ranae*), the water violet (*Hottonia palustris*), and the pond-weed (*Potamogeton natans*). On the surface of the wider expanses of stiller water one may see the white water-lily (*Nymphaea alba*), the yellow water-lily (*Nuphar luteum*), and the leaves of the smallest of our British flowering plants, the common floating duckweed (*Lemna minor*). In some of the marshy places are the great yellow blossoms and shining green leaves of the marsh-marigold, Mary-buds or king-cups (*Caltha palustris*), one of our most beautiful of English flowers; also the marshy fields may give us the cuckoo flower or lady's smock (*Cardamine pratensis*) with its pale lilac flowers.

(iv) The fourth distinctive region for plants adapted to their environment is that of the sand dunes and sand wastes along the coast, together with the shallow pools or "slacks" where water accumulates during the winter. Here in the hollows is often formed a thin deposit of rich soil very favourable for the growth of wild-flowers. Some of the best-known flowering plants found among these dunes are the rest-harrow (*Ononis arvensis*), with its pea-like pink blossoms; the sand convolvulus or small bindweed (*Convolvulus arvensis*), with white or pink funnel-shaped flowers; the scarlet pimpernel(*Anagallis arvensis*); and more rarely the large purple-flowered cranesbill (*Geranium sanguineum*) and various species of evening primrose (*Oenothera*).

Somewhat more rarely one sees the pretty yellow-centaury (*Chlora perfoliata*), one of the most beautiful of the gentian family, and sheltered under the dwarf willows which occur here and there may be found the pretty yellow and brown orchid (*Epipactis latifolia*). We should also note those plants which are the guardians of the sandhills, of which the most familiar is the sand marram grass (*Ammophila arundinacea*), the long underground stems of which thrust themselves through the sand and send up clusters of long rigid leaves. The stems and leaves of this and other grasses bind the sand together and retain it in position, thus preventing inland movement of the dunes.

(v) There is a fifth type of coastal vegetation which may be well seen to the north of Southport. Here is a long stretch of salt marsh washed at high tides. In early

June vast stretches of it are carpeted with the thrift or sea pink (*Armeria vulgaris*); and a month or two later the same ground is covered, though not quite so thickly, with the purplish blue sea-aster (*Aster Tripolium*). Here also, and in the similar but smaller stretches of salt marsh south of Formby, one may find the pink flowers of the sea spurrey (*Spergularia marina*), one of the relatives of the stitchwort.

10. People.

The earliest men who inhabited the British Isles were Palaeolithic men—that is, men of the Older Stone Age, who made rude stone weapons. It was formerly thought that they did not come so far north as Lancashire, but within the last few years a number of weapons and tools which were probably left by Palaeolithic man have been found in a river gravel at Hapton; and certain finds of worked flints on the Rossendale Fells have been regarded as being of latest Palaeolithic, or transitional between that Age and the succeeding Neolithic period (or Newer Stone Age).

The Neolithic men were more highly skilled and could make more effective weapons of stone and bone. They domesticated animals and grew cereals. The weapons and implements of these men are frequently found in South Lancashire; among other places at Clitheroe, Nelson, Burnley, Todmorden, Bacup, Bolton, Rochdale, and Old ham, generally on the moors above these towns.

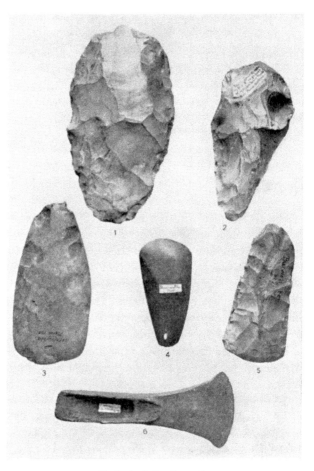

Prehistoric Implements

1, 2 *Palaeolithic;*   3, 4, 5 *Neolithic;*   6 *Bronze Age*

These Neolithic men were succeeded by a taller and more highly civilised people, who could make bronze weapons and had thus a very great advantage. They are supposed to have come in two waves of conquest. They do not seem to have exterminated or entirely driven out the earlier peoples, and it is thought that the descendants of Neolithic men survived in the remoter hill regions until a comparatively late period. The common traditions of "The Little Men among the Hills,"so frequent in North-east Lancashire seem to be connected with fugitive peoples who lived in the moorland and woodland fastnesses.

Later invaders had learnt the use of iron before the coming of the Romans; they had also learnt to make good pottery, beautiful weapons, and utensils, and were in many ways well advanced in civilisation. They buried their dead in mounds of earth or "barrows," some of which are found in South Lancashire. These Celtic-speaking peoples were in possession of the country when the Romans came; and some of them long survived in the moorland districts of South Lancashire, perhaps in Pendle Forest and on the Pennine slopes longest of all. Within the last hundred years the shepherds of the hills still counted their sheep in a sort of Celtic.

The Romans did not reach South Lancashire until the first century A.D. The famous Julius Agricola conquered the Welshmen at Mona, and then proceeded to subdue the Britons of the North-west. The Roman occupation of this area was largely a military one, and it had comparatively little effect upon the people or upon their language. A few Roman names have come down to us in South

Lancashire: the "streets" here and there, Colne (from Colonia or Colunio), Manchester (which is half Celtic, half Roman), and Ribchester, are examples.

Roman Altar, found in Manchester in 1831

The Anglo-Saxons entered South Lancashire in the sixth and seventh centuries by two routes. The Angles of the kingdom of Northumbria came through the Craven or Aire Gap and swarmed into the valley of the Ribble; the Anglo-Saxons of Mercia came in by the Midland Gate. Thus South Lancashire became in part attached to the kingdom of Northumbria, while a part belonged to Mercia.

The next invasion of South Lancashire was by Danes and Norsemen in the ninth and tenth centuries. Some came in from the north-east, by the Craven Gap, and entered by the openings at Colne and Clitheroe; others came from the north-west by sea and settled chiefly on the coast lands.

We have thus several waves of conquest after the Neolithic Period, and four groups of place-names are left as reminders of these successive peoples. The Celtic-speaking peoples left us many names of natural features: Pendle (at least the first part of it); the river names, Ribble, Calder, Don, Darwen (= Derwent), Douglas, Irwell, and Irk; the first part of the name Mamcestre, the first part of the names Cliderhow (now Clitheroe), and Calderstones, near Liverpool. Anglo-Saxon names are also common,—with the terminals ton, ham, burn, ley, leigh,—of which there are examples all over the county; and Norse names include *beck*, which is used regularly for a small stream in the east of the county, *kirk* (a church) which occurs in Ormskirk and Kirkdale, and *by* (a township) in Derby, Crosby, Formby, and Roby in the west of the county.

In the people of South Lancashire we thus find probable mixtures of Britons (the Celtic-speaking peoples), Anglo-Saxons, and Norsemen, with possibly some sprinkling of earlier Neolithic. This composite people lived, through the long Medieval period, in a region which was far removed from the richer and more fertile parts of Britain. This was more especially so in the moorland districts of the east, and that is probably the reason why there are still so many words and word-endings of Danish and Old English in use in the districts between Pendle and Boulsworth.

The people in South Lancashire were regarded as rude in the later Middle Ages and in early modern times. Witchcraft was implicitly believed in, and so late as the seventeenth century witches were executed. Pendle Forest was the most famous home of these reputed witches.

With the coming of the industrial revolution South Lancashire underwent a vast change. One of the most sparsely peopled regions of England two hundred years ago, it is now the most thickly inhabited outside London. The density of population is now about 3800 to the square mile, against an average density of 618 to the square mile for the whole of England and Wales. It is to be expected, therefore, that the population will be a "cosmopolitan" one, and there have been extensive "colonies" of workers from Cornwall, East Anglia, and other districts where wages were lower. The cities of Liverpool and Manchester have large Welsh and Irish colonies, and there are many Scotsmen in all the prosperous Lancashire towns. Manchester has a large number of Jews, with Jewish newspapers and synagogues. A strong Greek colony also exists in that city. Almost every country on the face of the earth is represented in the two great cities of South Lancashire.

11. History.

In this brief outline we can deal only with certain out-standing periods and stirring episodes.

The Roman Occupation. Tacitus tells us that the great Roman Governor Agricola began the wonderful series of Roman roads in the North. In the second century the Emperor Hadrian carried on the work of colonisation, and under him the first rampart was made across the country from the Tyne to the Solway. The Emperor Severus added a fosse or ditch and probably partly rebuilt and completed Hadrian's Wall, 208–211 A.D. Severus died at York, the Roman capital of Britain, in 211 A.D. From this time we may regard the conquest of northern England as complete.

The most important places of the North and North-west were Lincoln, Chester, York, and the forts at the ends of the Roman Wall, and the careful organisation of the Romans made roads to connect these places. For example, communication was necessary between Chester and York, and Manchester became an important place on the route. The great road over the Southern Pennines which connected Manchester with Ilkley and York is still to be seen on Blackstone edge, east of Littleborough. From Manchester a road ran across the western part of the Rossendale Fells to the fort at Ribchester on the Ribble, and another from Manchester crossed the east of the Rossendale Fells to the settlement at Colne, which was the centre of a number of vicinal or non-military roads. Another Roman road from Chester to the north crossed South Lancashire, through

Warrington and Wigan to Preston, and so northward to
Lancaster and Carlisle.

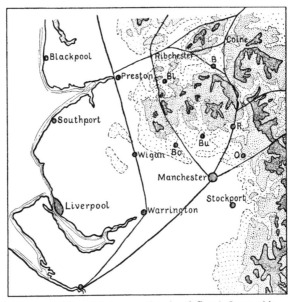

The probable chief Roman Roads of South Lancashire
and district, connecting principally Manchester (Mancunium), Colne (Colunio) and Ribchester (Bremetenacum)

*Land over 300 ft., 600 ft., and 1200 ft. stippled. The chief
modern cities and towns are shown*

From Ribchester a road ran through the Craven Gap
to Ilkley, Tadcaster, and York. Starting from the southern
bank of the Ribble at Ribchester it led east to Whalley,
where it crossed the Calder at Potter's Ford; it continued

past Clitheroe north-eastward to Chatburn and Downham, where it left South Lancashire. From near Whalley another road diverged to proceed along the southern slopes of the Pendle Range to the fort at Colne. From the latter place several roads seem to have radiated; one to the north to join the great road through the Craven Gap; a second to the east to pass along the northern slopes of Boulsworth towards Bradford and Wakefield; and another southward to join the great road over Blackstone Edge.

Many Roman relics have been discovered in South Lancashire, at Manchester, Warrington, Littleborough, Burnley, Colne, and Whalley; and, of course, at Ribchester, which is just north of the Ribble. A number of "camps" are seen on the moorlands of East Lancashire, which are usually spoken of as "Roman Camps"; and such town names as Stretford, Manchester, Colne, and Ribchester; together with Streetgate in Royton, Caster Cliff near Colne, Tum Hill (Latin *tumulus*) at Nelson, are all reminders of the Roman occupation, which came to an end in the early part of the fifth century.

There is very little trustworthy history of South Lancashire for the five or six centuries after the break up of the Roman Empire, and the little we have is so interwoven with tradition that it is not easy to separate the true from the legendary. The Anglo-Saxon conquest began about the middle of the fifth century, and it was two hundred years before the north was conquered. The resistance to the Anglo-Saxon conquerors was led for a time by the heroic King Arthur, but the history of his exploits is mixed with obvious legend. Geoffrey of Monmouth, who did not write until the middle of the twelfth century, is the doubtful authority for the history of King Arthur. He tells us that his hero won a famous victory over the Saxons near the river Douglas in Lancashire, about the year 500 A.D., and is said to have pursued the invaders through the Craven Gap towards York.

In the seventh century the two kingdoms of Mercia and Northumbria were rivals for the possession of the Lancashire Plain, and the tide of warfare ebbed and flowed through the Craven Gap and over the plain. Ethelfrith of Northumbria defeated the British at Chester in 613 A.D., and attached South Lancashire and Cheshire to Northumbria for the time being. In 642 A.D. the Northumbrian King Oswald was killed at Maserfield (possibly the Makerfield of to-day). There is a well in that district which is St Oswald's well to this day, and the church at Winwick is dedicated to St Oswald.

A church was founded at Whalley in the seventh century. There are strong traditions that the famous missionary Paulinus visited this district, and the crosses at Whalley and Burnley are popularly spoken of as the Crosses of Paulinus. Those at Whalley are almost certainly of a much later date; that at Burnley may possibly be of the eighth century.

The Anglo-Saxon Chronicle and the later chronicler, Simeon of Durham, tell of a great rebel leader called Wada, who was defeated by Eardulf, King of Northumbria, in a battle at Hwalleage in 798 A.D. Simeon puts the scene of the battle at Billingahoth near Walalaga. There are two villages, Billington and Langho near Whalley, and it is probable that the battle was fought here.

One other great episode in these dim times concerned South Lancashire. At the end of the ninth century Mercia had been overcome and to a great extent absorbed by Wessex; and Northumbria, though still strong, had become largely Danish. The Northern kingdom formed alliances with the Scottish and Northern Irish, and sought their help in the constant wars of the time. Edward the Elder and Athelstan, son and grandson respectively of the great King Alfred, had wrested West Lancashire and Ribblesdale from Northumbria. Whalley was thus transferred from the diocese of York to that of Lichfield, to which it belonged for hundreds of years. In connection with the rivalry between the two great kingdoms there was fought the great battle of Brunanburh in 937 A.D. The Northern kingdom, with its Danish and Scottish allies, was decisively beaten and Wessex became supreme. A famous epic poem, descriptive of the battle, finds a place in the Anglo-Saxon Chronicle.

There have been endless disputes about the locality of Brunanburh, and one set of authorities claims the moors south-east of Burnley as the site. Extensive earthworks are found there, and the name may have had to do with the river Brun, the river on which Burnley stands. A great find of coins was made at Cuerdale near Preston in 1840; this seems to have been the treasure chest of a Northern or Danish king, and it had been buried or lost some time after 910 A.D., the latest date on any of the coins. It is suggested that a king of Strathclyde, hard pressed after the battle, was retreating to the north, and either lost or buried the chest in his flight. But the mystery of Brunanburh cannot be considered as fully solved.

The Later Middle Ages. There is little to relate of the history of South Lancashire during the centuries between the Norman Conquest and the dawn of the Modern Period, but the founding of the great religious houses was an important phase in the history of those times, and South Lancashire saw something of that movement. There was

Whalley Abbey

a Cistercian Abbey on the low-lying land at Stanlaw in Cheshire, and the monks, afraid that the inroads of the sea would ultimately engulf their Abbey, migrated to the flat meadows by the Lancashire Calder, where the Whalley Church had existed for some hundreds of years. Whalley Abbey was founded in 1296; the consecration took place in 1306. For two and a half centuries the monks of Whalley lived their lives here. The Abbot took part in the Pilgrimage of Grace in 1536, and on the suppression of this "insurrection" he was taken, first to Lancaster and then to Whalley, where on the 12th of March 1537 he was executed on Clerk Hill in sight of his own great monastery.

Priories were also founded at Burscough and Upholland. There were also colleges of monks, one of which was at Eccles[1]; and the old grange at Eccles still bears the name of Monks' Hall. During this period there were founded "Chantry Schools" where a part of the priests' duty was keeping a school; such "Schools" existed at Manchester, Warrington, Middleton, Leyland, Blackburn, Clitheroe, Burnley, and probably at Colne. These schools were the forerunners of the grammar schools of later days.

The two great orders of Friars had their quarters or "houses" in the towns, the Franciscans or Grey Friars at Warrington, and the Dominicans or Black Friars at Manchester.

Of the Wars of the Roses South Lancashire saw but little. Henry VI, the Lancastrian king, was in hiding almost on the border of Lancashire in 1464–5, after the defeat at Hexham. Bolton Hall is in the West Riding of Yorkshire, three miles from the boundary, and here the fugitive was hidden for several months. He was betrayed, and captured by agents of the House of York at the "Hipping Stones" at Brungerley, Clitheroe, and taken from there to London.

In the Civil Wars, one of the earliest battles of the north was fought at Read Lane, east of Whalley, in 1643.

---

[1] Latin—*ecclesia*, a church.

The Earl of Derby had set up the King's standard and defeated the Parliamentary forces at Warrington on April 5th, 1643. The Earl decided to strike through the Pennines at the strong Parliamentary forces on the eastern side, and he therefore occupied Whalley. Colonel Shuttleworth of Gawthorpe Hall hastily collected what forces he could to oppose the Earl's advance, and a battle ensued about two

Brungerley Bridge, Clitheroe

miles east of Whalley for the control of the Roman Road to Colne. The Royalists were defeated, and driven back on Whalley. There were other skirmishes at Colne and Clitheroe in July 1643, in both of which the Royalists had the worst of it, and the Craven Gap remained in the hands of the Parliamentarians.

The year 1644 was critical. The Earl of Newcastle, the Royalist leader in the North, was practically shut up in

York; and to Oxford, where were Charles and his nephew Prince Rupert, there came urgent appeals for help from York, and from the Countess of Derby who was holding Lathom House against a strong besieging force. Rupert set out to relieve the north. He reached Lancashire and soon won the battle of Bolton; showing, it was said, no mercy to the defeated Parliamentary forces. He then crossed the plain to Liverpool, which he relieved, thence going northward to the succour of Lathom House. The siege of the historic mansion was soon raised, and Rupert hurried on through the Craven Gap to the relief of York. His advance guard was at Skipton on the 27th of June; the bulk of his army crossed out of Lancashire by Clitheroe and Colne on the 29th. Then followed the decisive battle of Marston Moor, near York, on July 2nd, and Rupert's victorious career was checked. Prince Rupert retired through the Pennines by the north-western or Wenning opening, as Shuttleworth and Assheton held the direct route to South Lancashire.

In 1648 South Lancashire saw another phase of this great conflict. The Scottish Presbyterians were now on the King's side, and an army marched into England to help the Royalists, choosing the route through Westmorland and Lancashire in preference to the one through the north-eastern plain, as it was confidently expected there would be more help for the King on the west side of the Pennines. In August 1648, Hamilton marched his forces towards Preston, and Cromwell hurried through the Craven Gap from the east. On the 15th August Cromwell's forces were between Clitheroe and Gisburn. On the 16th a Council of

War was held at Clitheroe; the Parliamentary army crossed the Ribble, and marched towards Preston. The armies met near the latter town and Cromwell was again victorious. The victor followed the defeated army into South Lancashire, and two minor engagements were fought at Wigan and Warrington, both of which added to the misfortunes

Clitheroe Castle

of the King's cause. It was from Warrington that Cromwell wrote his famous letters to the Speaker of the House of Commons.

There is still the campaign of 1651. Prince Charles (Charles II) invaded England, again on the western side. On his march southward he entered South Lancashire and there was a skirmish with General Lambert at Warrington. Lambert did not succeed in stopping the southward progress of the Royalists, who continued their march to the south-west, only to be met by Cromwell's forces at Worcester. There the army of Prince Charles was finally defeated, and the Civil War was at an end.

The Jacobite Risings of 1715 and 1745.

South Lancashire had but little share in the affair of 1715. The leaders of the Stuart forces chose the western side of the Pennines again for their march into England, expecting that the numerous Roman Catholics of Lancashire would rise in support of the Stuart cause. The comparatively small rebel army marched through Lancaster and on to Preston. The leaders hoped to get to Warrington, and so to control all South Lancashire; but they were caught at Preston and there they surrendered. "The '15" was soon over.

The 1745 rebellion was more serious. After some successes in Scotland, the Stuart army marched into England, following much the same route that the 1715 army had followed. There was no hostile army at Preston this time and the invaders entered South Lancashire and marched to Manchester. Here a stay of some days was made and over two hundred men were enlisted to form the "Manchester Regiment," with Francis Towneley of Burnley as Colonel. But South Lancashire did not rise as had been expected, and Liverpool took active measures against the Stuarts, a regiment called "The Liverpool Blues" being raised to support the Government. The invading force went on to Derby, but there was still no sign of a general rising, and the disastrous retreat was commenced. So far as South Lancashire was concerned the 1745 rising was over.

Meanwhile the *Industrial Revolution* was beginning, and the next hundred years were to change the whole face of South Lancashire and to witness the marvellous growth of its cotton, chemical, and engineering industries, and make it the most densely peopled of provincial counties.

Most of the great inventions which brought about the growth of the textile manufactures were of South Lancashire origin. Up to 1738 all spinning of yarn and weaving of cloth was done by hand. In 1739 Kay of Bury invented the "flying shuttle" which at once revolutionised the speed of weaving, and left the spinners far behind. Twenty years later his son invented the drop-box for shuttles, which made it possible to weave with various colours of weft without taking out the shuttles by hand. Hargreaves of Blackburn invented the "Spinning Jenny" about 1766, and made it possible to produce yarn at a much greater rate than before. A little later Arkwright of Preston and Bolton invented a more complex and effective machine which improved the spinning branch of the cotton industry still further; and later still Crompton of Bolton, by the invention of his wonderful "mule" spinning machine, brought that particular series of inventions to a close. Spinning had now completely outdistanced weaving; but the required invention of a new and more effective loom was not long in coming. This was the work of Cartwright, a Derbyshire clergyman. All this had taken place between 1739 and 1785; and as James Watt had meanwhile improved the steam engine almost out of recognition, and furnaces could now turn out iron at a rate which was marvellous compared with anything previously known, the textile and allied industries developed very greatly.

In the next sixty years there were further minor improvements in spinning machinery; the Power-Loom was further improved by Roberts of Manchester and others into a more effective machine; calico printing in the hands of the Peels made great strides; bleaching was more rapidly done by using chlorine; dyeing and finishing were made to produce wonderful results, and by the time of the Reform Era (1830–1840) South Lancashire had become the greatest hive of industry the world had till then known; and the rapid and uncontrolled growth of the towns had induced serious problems of housing, sanitation, education, and social outlook which are not even yet solved.

About this time Mercer of Great Harwood had invented his famous process for the mercerisation of cotton; and somewhat later Lightfoot of Accrington had made the first aniline black. By 1870 the specialisation of the cotton industry had adjusted itself broadly as we see it to-day, and the progress since that time has been mainly in the cheapening of production, and in the adaptation of the industry to finer and more expensive goods. The development of transport by canal, road, and railway during this great industrial epoch is considered later.

12. Industries.

The first of all industries in any country is that of Agriculture. Perhaps we should not expect this industry to rank high in South Lancashire, but the western plain is really a most important agricultural region. Experimental work of great importance is done at the Potato Testing Station of the National Institute of Agricultural Botany, Ormskirk, and also at the Lancashire County Council's Farm and Garden at Hutton.

The peaty soil of the Lancashire Plain seems to be particularly suitable for potatoes and vast quantities are grown, to be marketed at Ormskirk, which is one of the most important potato markets in the British Isles. In travelling across the western plain from Wigan to Southport or from Liverpool to Preston the observer sees vast fields of potatoes and cabbages, and acres of celery and rhubarb. The whole of Lancashire, North and South, had 40,729 acres under potatoes in 1926, out of a total of 469,655 acres for England and Wales; that is, Lancashire had above one-twelfth of the total acreage under potatoes, though its area is less than one-thirtieth of the whole country. Oats are produced in considerable quantity, but not much barley and wheat.

The production of milk and butter is naturally very important as there is such a large demand by the immense population. A good deal has, however, to be imported from Cheshire and Derbyshire on the one hand, and from Craven and Bowland on the other. Lancashire, as a whole, had 222,830 cattle in 1926, which is a high proportion for a county with such a large share of moorland. The Lancashire Plain produces an excellent type of cheese, known as "Lancashire cheese," which is largely marketed at Preston and Liverpool.

Quarrying and Mining.

Big quarries have been opened out in the massive lime-stone beds of the Carboniferous series at Clitheroe and Chatburn. The dark-coloured, harder limestone is much used for road-metal; from the light grey limestone lime is made, and large quantities of limestone and lime are sent away to the ironworks at Darwen and Wigan to be used as a flux in iron smelting. Some cement is also made from the limestone in the Clitheroe district.

Above the Carboniferous Limestone series comes the Millstone Grit. This is quarried at many places along the Pendle range of hills, notably at Colne and Wiswell near Whalley. In the Rossendale district the Millstone Grit is quarried as a building stone, and there are many quarries where the famous Rossendale Flags and Haslingden Flags are obtained. There are numerous quarries on the moors between Rochdale and Bacup.

The Coal Measure Sandstones are quarried at many points all round the Rossendale Fells, where the Lower Coal Measures are well exposed, as at Nelson, Burnley, Accrington, Darwen, Rochdale, Littleborough, Royton, and Oldham. Stone buildings are a notable feature in most East Lancashire towns. From the Coal Measures, shale for brick-making is dug at Nelson, Burnley, Accrington, Darwen, Chorley, Wigan, St Helens, Droylsden, and Bolton. Accrington district has become famous for the manufacture of various kinds of bricks, tiles, and drain-pipes; and Burnley specialises in sanitary ware.

Coal mining is carried on in almost all parts of the

Lancashire coal-field, some centres of the industry being Wigan, Burnley, and Bolton. There are many famous coal seams of which the Lower Mountain Mine or Ganister Mine, the Arley, the Orrell Five Feet, the Wigan Nine Feet, and the Great or Bing Mine are perhaps the best known. The Arley seam is at the base of the Middle Coal Measures and is known as the Little Delf seam at St Helens, the Royley Mine at Rochdale and Oldham, the Habergham Mine, and the Marsden Four Foot in the Burnley Basin. The famous Wigan cannel seam is now nearly exhausted; and in the Burnley Basin the Burnley Four Foot and the Doghole seams are practically worked out.

The Lancashire coal-field is in the second rank of British coal-fields, being surpassed in yield of coal and number of persons employed by the Northumberland and Durham; the Yorkshire, Derby, and Nottingham; the South Wales; and the Lanark coal-fields. There were 105,000 persons employed in the South Lancashire coal-field, in 1924, and the production was 19,540,000 tons of coal.

Smelting-coke. There is a constant demand for coke for iron-smelting, and ordinary gas coke is not suitable for the purpose. Hence special ovens are constructed where coal is heated to convert it into smelting-coke; this industry is carried on in North-east Lancashire (Burnley and Accrington), and at Upholland and Wigan chiefly. There are two types of oven in use, one where the tar and ammonia are collected, and the other where no attempt is made to do so. South Lancashire had 862 coke ovens of various kinds at work in 1918, and was the fifth county in Great Britain in this respect, coming after Durham, Yorkshire, Glamorgan, and Monmouth.

Finally, it should be mentioned that there is a considerable industry in peat-cutting on some of the "Mosses" to the west of Manchester, and the traveller by rail from Manchester to Liverpool will notice quantities of peat being dried mainly for fuel and litter for horses.

The Chemical Industries.

We may conveniently divide the chemical industries into two great groups—Primary Chemicals and Derived Chemicals, the former being the raw materials used in the manufacture of the latter. The Primary Chemicals—likewise for convenience—may be again divided into two sub-groups, though there is no absolute dividing line between the two. They are the Common Salt chemicals and the Coal Tar chemicals.

The Common Salt Group. This branch of chemical manufacture is chiefly located in the Mersey region owing to the convenience of its situation. The salt is obtained from Cheshire; the fuel comes from the South Lancashire coal-field; while the great Mersey estuary, the Manchester Ship Canal, the Bridgewater and other barge canals, and a vast network of railways, afford abundant transport facilities.

Common salt is a compound of two elements, sodium and chlorine, and these may help us to classify the chief chemicals which are made in that region. The sodium group includes caustic soda, soda ash, washing soda,

bicarbonate of soda, sodium sulphide, and sodium hyposulphite; the chlorine group includes chlorine itself,
bleaching powder, sodium hypochlorite (with "chloros"
and "chlorazone"), chlorate of potash, chlorate of soda,
and hydrochloric acid. Attached to this group is the

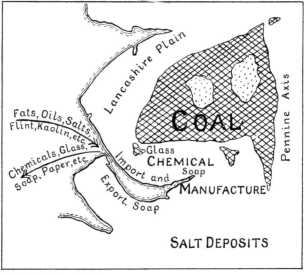

The relation of the Chemical Industries of the Mersey region to
the fundamental factors: raw material, fuel, transport, markets

manufacture of the two important mineral acids—Sulphuric Acid and Nitric Acid. South Lancashire makes
more of these than any region in the world, and is also a
great consumer.

It will be noticed that this group of chemicals includes
the chief acids and alkalies of commerce, and the whole

group is often spoken of as "The Heavy Chemicals." Along a belt starting from the suburbs of Liverpool to St Helens, through Garston and Widnes to Warrington, and at intervals eastward to Manchester, the manufacture of these chemicals is carried on. Widnes may be regarded as the chief centre, being almost entirely given over to this and closely allied industries. There are important works in other parts of industrial South Lancashire, among which may be mentioned those at Church and Hapton in North-east Lancashire. The crude ammonia from the coal industries finds its way into this great group. Compounds of ammonia with hydrochloric or sulphuric acid are of very great importance.

The other sub-group of Primary Chemicals starts from Coal Tar and Coke Tar as its primary raw material. Tar is a substance of very variable composition, available in enormous quantities in South Lancashire. It is distilled in different ways, the precise method varying according to the nature of the tar, and the product desired. A series of "oils" distil over, which are condensed at different temperatures, and from these, in their turn, a wonderful variety of useful products are made, of which it would be quite impossible to give an exhaustive list here, though among them may be mentioned: benzene, toluene, naphthalene; carbolic acid, cresylic acid, salicylic acid, pyrogallic acid, hydroquinone; picric acid, trinitro-toluene ("T.N.T."), nitrobenzene, aniline; many wonderful dyes, among which is artificial indigo; many perfumes, colouring and flavouring matters; and a long list of important medicinal substances. Among other chemicals which are made in South Lancashire may be mentioned—sulphur, permanganate of potash, permanganate of soda, potassium bichromate, lead acetate, zinc sulphate, salts of tin, compounds of arsenic, and superphosphate of lime. The list might be extended to a great length.

The second great group is that of *Derived Chemicals*; here the Primary Chemicals of the first great group become raw materials; of this group we will mention five classes of substances only: soap, paper, leather, glass, and paints.

Soap. The raw materials for this manufacture are caustic soda and vegetable or animal fats and oils, such as tallow, palm oil, cotton-seed oil, and rosin; hence it is carried on in the Mersey region, and Widnes and Warrington are the homes of soap works, the names of whose brands of soap are household words. There are many soap works scattered here and there in South Lancashire, of which those at Irlam and Clayton-le-Moors may be mentioned.

Paper. The raw materials for this are wood pulp, esparto grass, alfalfa, cotton and linen rags, and alkalies. Starch, resin, alum, china-clay, and barytes are also used in varying degree for different classes of paper. Very important indeed is an abundant supply of pure soft water. It will readily be understood why the western slopes of the Pennines, the edges of the Rossendale Fells, and the lower valley of the Irwell have paper-mills almost everywhere. Burnley, Accrington, Darwen, Chorley, Bolton, Ramsbottom, Haslingden, Heywood, Rochdale, Radcliffe, Middleton, and the suburbs of Manchester are towns where paper is made. Bury has been styled the "Metropolis of Paper Manufacture," as it not only makes large quantities of paper, but is also famous for its production of paper-making machinery.

Leather. This manufacture needs lime, soda, sodium sulphide, sulphuric acid, hydrochloric acid, ammonia, sodium hyposulphite, potassium bichromate, and chrome alum, together with many kinds of vegetable tanning materials. There are also, of course, the skins of sheep, goats, cattle, and horses. Most of the necessary chemicals are produced in South Lancashire; the vegetable tanning substances come from many parts of the world, and in addition to home supplies there is a big import of hides and skins. A liberal supply of good water is absolutely necessary, and the Pennine slopes and Rossendale Fells supply this. It is thus not difficult to understand why there are leather works at Colne, Burnley, Bury, Bolton, Chorley, Rochdale, Bacup, Heywood, Middleton, Oldham, Manchester, Warrington, and Liverpool.

Glass requires fine sand or powdered flint, together with soda ash, salt cake, lime, and smaller quantities of other chemicals; a liberal supply of coal as fuel is also needed. This industry is localised chiefly at St Helens and district, but there are glass-works at Newton-le-Willows and Warrington, and smaller works in some of the larger towns.

Paints and Varnishes. The raw materials for this group are linseed oil, rosin, turpentine, shellac and other gums, with lead acetate, red lead, zinc oxide, barytes, lead chromate, and iron oxides. Liverpool, Manchester, Widnes, Bolton, and Darwen are the chief places where this group of derived chemicals is made.

It will be noted that important factors for the success of all these derived chemical industries are cheap fuel, cheap and efficient transport (home and abroad), and the presence of a great consuming public. Large quantities of all of them are exported, and the Mersey estuary and the Ship Canal are clearly important factors in this connexion.

Though no attempt can be made here to include all the many forms of industry connected with chemicals in the county, some comparatively recent introductions may be mentioned. The Rubber trade has now become very important, and there are rubber works at Leyland, Warrington, Rochdale, Littleborough, and Manchester. The waterproofing trade uses a great quantity of rubber, and this is a considerable industry in Manchester, Warrington, and Leyland. Another industry is the manufacture of various forms of asbestos products; this is carried on at Widnes, Manchester, Ashton-under-Lyne, and in the Rochdale district including Littleborough. Rectified spirit, methylated spirit, acetone, and formaldehyde form another group, the manufacture of which is carried on at Liverpool, Manchester, and Bury.

Finally, the extraction of copper from copper pyrites is now an industry closely associated with that of the mineral acids, and is done at Widnes and at Hapton. Copper smelting is carried on at Widnes.

Iron and Engineering.

There is little iron ore obtained in South Lancashire, hence iron smelting is not an important industry. In spite of this there is a great engineering industry, and iron foundries (as distinct from iron furnaces) are numerous. Manchester indeed, with its suburbs, and the towns quite close to it ("Greater Manchester") has been described as the greatest engineering region in the world.

Textile engineering is one of the chief branches, and almost the whole of the highly specialised machinery used in the cotton trade is made in South Lancashire, and the adjacent parts of Cheshire and Derbyshire. Almost every town contributes its share, and some even of the smaller towns manufacture one speciality or another for use in that highly developed industry. Textile engineering on the great scale is carried on in Manchester and Salford, Ashton-under-Lyne, Oldham, Rochdale, Bury, Bolton, Blackburn, Accrington, and Burnley. Probably the greatest works of this kind are in Oldham; there are big works at Bolton, and at Accrington there is one which employs 4000 to 5000 people. Burnley has no very big works, but boasts a considerable number of smaller ones, and claims to turn out more looms than any other town in the world.

Steam boilers are made at many places. Greater Manchester makes steam engines, oil engines, gas engines, electric motors, dynamos, electric fans, pumps, boring machinery, presses, crushing and grinding machinery, lathes, drills, laundry, printing, and fruit-cleaning machinery; and its great engineering works on the banks of the Ship Canal are among the largest in England. Motor cars, motor wagons, and aeroplane parts are also made in the Manchester district.

Outside the immediate range of Manchester there are engineering works of many other kinds scattered over South Lancashire. Motor wagons are made at Leyland;

the London, Midland and Scottish Railway builds power-
ful railway engines suited to high gradients at Horwich;
and railway wagons are made at many places. The famous
Nasmyth steam hammers come from Eccles; crushing
and grinding machines are made in the St Helens and
Widnes district; and flour-milling machinery at Liverpool,
Manchester, and Rochdale. Bury specialises in machinery
for paper making; Bootle and Liverpool make marine
engines and ships' fittings of various kinds; Prescot makes
electric cables and has recently revived its old watch-
making industry; Warrington specialises in many kinds
of wire; Ashton-in-Makerfield makes nuts, files, and
locks; Atherton nails, bolts, and spindles; Burnley over-
head "runways," and machinery for crushing oil nuts,
bones and sanitary refuse; Colne iron cranes and machinery for cleaning dried fruit; and so one might con-
tinue until the list would almost have the appearance of a
Trade Directory. Enough has been given to show the
magnitude of this great industry, and the wonderful variety
of things made.

The Textile Industries.

The South Lancashire industries which we have discussed up to this point are very important, and in most
parts of the world would be considered as of the first
rank. Here they take second place in comparison with the
greatest of all, the manufacture of cotton. This great
industry, alike in the number of people employed, in the
value of goods produced, in the wonderful perfection of
its organisation, and in the skill displayed throughout, is

undoubtedly the premier manufacturing industry of the British Isles, and is probably the one by which we are best known abroad.

It is all the more wonderful when we consider that all the raw cotton has to be imported from tropical and sub-tropical lands, and that three-quarters of the goods manufactured are for export. The value of the exports of cotton yarn and cotton piece goods (including cotton thread and cotton lace) in 1924 amounted to the colossal sum of about £200,000,000. This was nearly as much as the value of the combined exports of woollens and worsteds of all kinds, clothing, iron and steel manufactures and machinery.

The industry is almost entirely confined to the eastern half of South Lancashire together with the immediately adjoining parts of North Lancashire, the West Riding of Yorkshire, Derbyshire, and Cheshire. This we may regard as the Manchester province. There are four chief groups of processes of manufacture in the industry—Spinning, Weaving, special applications of Chemistry, and the Cotton Waste Trade. We shall discuss the distribution of these in South Lancashire.

The *Spinning* branch is located in South-east Lancashire, in Greater Manchester and a crescentic girdle of towns not more than 12 to 15 miles away. These towns are—Stockport (in Cheshire, but with suburbs in South Lancashire), Ashton-under-Lyne, "Greater Oldham," Rochdale, Heywood, Bolton, Wigan, Eccles, and Leigh. Within this crescent of towns there is internal specialisation. Oldham district spins chiefly middle-grade American cotton; Bolton Egyptian and high-grade American;

Stockport and Rochdale do much doubling, that is twisting together two or three threads of fine yarn into one strong "doubled" yarn. In this spinning region there are about four-fifths of the total of over 40,000,000 spindles in the Manchester province.

Spinning

The other great operation is *Weaving*, which is mainly localised in the Calder and Ribble valleys to the north of the Rossendale Fells. Here are found about three-quarters of the looms in the Manchester province, but only about 15 per cent. of the spindles. The chief towns in this weaving zone are Colne, Nelson, Burnley, Accrington, Blackburn, Darwen, Clitheroe, Chorley, and Preston (which is mainly north of the Ribble and is therefore in North Lancashire). In this weaving belt are some of the finest weaving mills in the world, and the variety of cotton cloths produced is amazing, from cheap Burnley "Printers" to high-priced blouse materials. The greatest weaving towns are Blackburn and Burnley, the former claiming to have more looms, and the latter to turn out more yards of

Weaving

cloth, than any other town in the world. Nelson is noted for the excellence of the grey cotton cloths produced and for the high earnings of its operatives; Colne, its neighbour, for the great variety of cotton goods woven there.

The third group of processes, the *application of Chemistry* to the Cotton Trade, includes bleaching, dyeing, printing, mercerising, and "finishing." These processes are carried out, as a rule, where there is a good and reliable water-supply, and make great demands on the chemical trades of the region. There are works having associations with historic names in the industry, some of which are of world-wide fame.

The fourth branch is the *Cotton Waste Trade*. Waste is, of course, a relative term, and in the keen competition

Dyeing in the Hank or Skein

of modern times little must be absolutely wasted. At every stage of the industry, from the opening of the bales to the final bundling of the cloth there is a certain amount of "waste." It is said that of a bale of cotton of 480 lbs. some 160 lbs., or one-third of the whole, finds its way into the hands of the Cotton Waste Trade. This branch is equipped with complex machinery to deal with the various types and

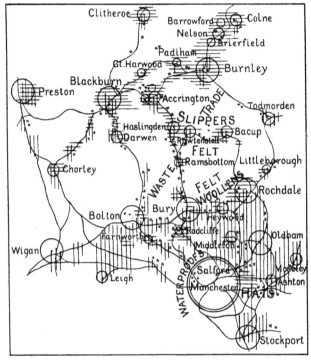

**Map showing the distribution of the Textile and allied
industries of South Lancashire**

Horizontal lines, cotton weaving. Vertical lines, cotton spinning
××× *Application of chemistry to textiles*

*The railways connecting the cotton industry areas
with Manchester are shown*

grades of waste; some of it is made into short-staple yarn, and is then woven into sheetings in the towns and villages of the Rossendale Fells; lamp and candle wicks, various kinds of cleaning cloths and "sponge cloths," and cotton flocks are other productions from it. The working up of Cotton Waste is carried on in many parts of the great cotton province, but especially in Manchester, Bolton, Burnley, and the towns of Rossendale.

Other Textile Industries.

The Cotton manufacture is by far the greatest of the South Lancashire industries; but the share of the county in other Textile industries is by no means negligible.

The Woollen industry has been carried on in Rochdale for centuries. As early as 1558 the flannel manufacture was accounted the staple trade of the town. To-day there are some thirty firms engaged in the flannel industry in Rochdale and district. Bury has also a share in this industry, and this town and Rossendale Valley are engaged in the allied work of making felt.

The manufacture of silk is carried on to some extent in Manchester, Oldham, and the district. Associated with it is the weaving of velvet, plush, and other pile fabrics; this is an industry of Oldham and its neighbourhood, where similar fabrics are now largely made of cotton. The manufacture of artificial or cellulose silk is also a South-east Lancashire industry.

Incandescent gas mantles are made of ramie in the neighbourhood of Manchester. There is a very considerable output of rope and twine, canvas bags, and sacks in

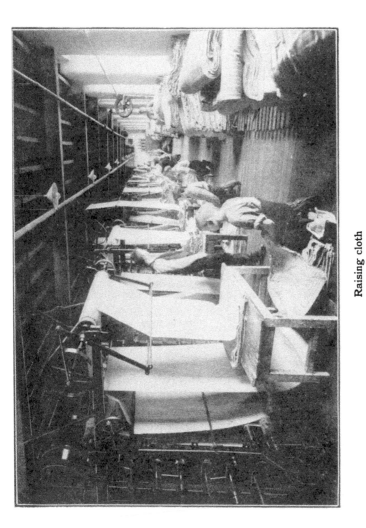

Raising cloth

A finishing process in the manufacture of cotton cloth, which produces an appearance and feel of fullness

Liverpool and Manchester, jute, hemp, and ramie being used as the chief raw materials, though cellulose fibre or paper fibre is now being employed for these purposes.

The Clothing Group of Industries.

In a great industrial region where a dense population provides a large market there is sure to be some development of the clothing industries. In South Lancashire these are increased by the abundant supply of textiles, and by association with the chemical trades of the region. Some of these manufactures are of comparatively recent growth, but they already provide employment for some thousands of people.

Silk hats and felt hats are made in the south-eastern corner of the county, where Denton is the centre of the industry. Cloth caps and ready-made clothing are turned out in large quantities both in Manchester and Liverpool. Underclothing, hosiery, and blouses are made on the whole-sale factory scale in Manchester, Wigan, Bolton, and Burnley.

The manufacture of slippers is now a big industry in Rawtenstall, Waterfoot, Newchurch, and Bacup, and some of the other villages of the Rossendale district. This claims to be the greatest slipper-making district in the world. In the original manufacture of Rossendale slippers felt was almost exclusively used for the uppers, and this clearly connects the industry with the felt industry of Bury and Rochdale, which is now firmly rooted in Rossendale as well. Leather is used for uppers as well as for soles; the heavier sole leather is produced at Bacup, Bolton, Bury, and Manchester, and a good deal of the light leather is tanned, dyed, and finished at Colne.

The manufacture of waterproof clothing as an industry has increased rapidly within quite recent years. Special cloths, woven in various parts of the cotton province, are made waterproof by a rubber-naphtha paste which is applied in very thin layers by machinery. This is now a large industry in Ardwick, Cheetham, Salford, and Pendleton, where many, waterproof garments of different types are produced. Raincoats are made in greater Manchester from fine cloth woven in Lancashire and the West Riding of Yorkshire. This is put through a special process known as chemical proofing before it is made up into garments.

13. Famous and Historic Buildings.

South Lancashire is not rich in ancient buildings, but there has grown up in the county an architecture connected with science, art, music, education, industry, and commerce which is of considerable importance.

Castles.

The only Norman castle of which any considerable part is left is that of Clitheroe. This is not mentioned in Domesday Book ; it was probably built by one of the De Lacy family a few years after 1086. It was dismantled in 1649, and fell into ruin very rapidly until only the keep and parts of the old walls remain. The castles at Liverpool and Bury mentioned by Leland the antiquary have disappeared long ago. There is still a Castle Street in Liverpool.

Monasteries and Priories.

There are the remains of three medieval religious houses in South Lancashire. The most considerable are those of the Cistercian Abbey of Whalley, founded in 1296. Two well-preserved gateways as well as other considerable portions of the ancient buildings exist. The doorway which led from the Chapter House is particularly fine; it is flanked by two pointed windows. Within the last few years excavations have been carried out and further parts of the old walls and bases of pillars have been uncovered. The tiled floor of the church has also been reached.

At Burscough, about two miles from Ormskirk, was a priory founded in the reign of Richard I by Robert of Lathom, whose seat was Lathom House. The Stanleys afterwards acquired Lathom House by marriage, and the members of that historic family were buried in Burscough Priory until the Dissolution of the Monasteries. The central piers of the priory are still left, a lonely ruin in the wide plain.

Upholland stands on rising ground west of the Douglas valley, and here was a Benedictine Priory, the remains of which may still be seen.

Historic Houses.

There are a good number of these in South Lancashire. Near the town of Prescot is the private park within which stands Knowsley Hall, the principal seat of the Earl of Derby, a large mansion "but not noteworthy by its architectural style."

Lathom House near Ormskirk played an important part in the Civil Wars. The old house was destroyed as the result of that war and it is doubtful if the modern Lathom House is even on the same site. Hoghton Tower near Blackburn is held by many to be one of the finest Renaissance mansions in the country, in spite of its having suffered

Hoghton Tower

much during the Civil War. Gawthorpe Hall near Padiham is one of the most typical Elizabethan houses in the North; it is the seat of Lord Shuttleworth. Towneley Hall near Burnley, for many centuries the home of the famous family of Towneley, is now a Corporation Museum and Art Gallery. Holme near Burnley, Emmott Hall near Colne, Hopwood Hall near Rochdale, Speke Hall and Croxteth

The Holme, Burnley

Hall-i'-th'-Wood Museum

Hall near Liverpool, Hale Hall near Warrington, and Formby Hall may be mentioned as other examples of famous houses. Hall-i'-th'-Wood near Bolton is a fine example of timbered architecture which was presented to the Corporation of Bolton by the first Lord Leverhulme. Roughlee Hall near Nelson, a long, low, gabled house, dating from 1536, was probably the home of the "Mistress Nutter" of Harrison Ainsworth's *Lancashire Witches*.

Cathedrals.

The bishoprics in South Lancashire are only modern and it cannot therefore boast of any ancient cathedrals. Manchester Cathedral, as restored in 1882, attempts to reproduce faithfully a fifteenth century church which had grown up on the site of a Saxon church. The choir stalls, erected in 1485, are claimed to be among the most beautiful in Europe. Liverpool has no ancient cathedral, but the fine one on St James's Mount now nearing completion will be one of the grandest in Europe, and the largest ecclesiastical building in England.

Churches.

There are many fine churches in the wealthy South Lancashire towns, many of them modern, but often on the site of earlier churches. It would almost seem that in building beautiful modern churches or restoring and rebuilding ancient ones, South Lancashire has been attempting to counterbalance the disfigurement of the countryside by its industrial buildings. The parish churches of Oldham, Rochdale, Bury, Bolton, Blackburn, and Burnley are examples, especially St Peter's, Burnley, which

Parish Church, Rochdale

Parish Church, Middleton

is placed on a fine river terrace above the Brun. It was originally built in the fourteenth century, but has been extensively modernised. The tower was erected about 1480; the Stansfield chantry dates from about 1310, and the Towneley chantry from 1372.

Among well-known churches in the lesser boroughs and rural districts Whalley parish church is noteworthy for its famous Runic crosses and other monuments. Middleton church has a Norman arch of about 1120, and tower and porch of the early fifteenth century. The church at Ormskirk has traces of Norman work, but is chiefly remarkable for having both a tower and a spire. It is supposed that the spire was built first, and the tower erected later to contain the bells which were moved from the Priory of Burscough. Other interesting old churches are Sefton, Upholland, Leyland, Colne, Eccles, and Ashton-under-Lyne. It is, however, in its modern buildings connected with education, art, science, etc., that South Lancashire chiefly excels. St George's Hall, Liverpool, is one of the finest modern buildings in England. It was designed on the lines of a Greek temple, and with its Corinthian colonnade is remarkably impressive. The Walker Art Gallery, the Picton Library and Museum, the Philharmonic Hall in Liverpool, the Rylands Library, the Free Trade Hall, and the Art Gallery in Manchester, the New Public Halls and Sessions House in Blackburn, the Lyceum and the Art Gallery and Museum at Oldham, the Museum and Library at Warrington, the Athenaeum and the Art Gallery at Bury —all these and many others are buildings characteristic of the spirit of modern South Lancashire.

South Porch, Parish Church, Middleton

City Halls and Municipal Buildings.

The great municipalities of South Lancashire have housed themselves in many fine town halls. Some of these are in the very front rank of such buildings. Manchester Town Hall is often claimed as the most striking in the kingdom; other notable ones are those of Liverpool, Bol-

Old Stocks, Colne Parish Churchyard

ton, Blackburn, Oldham, Rochdale, Bury, and Bootle. The beautiful Town Hall of Southport has the advantage of its setting in Lord Street, one of the best streets in the north of England.

Manchester and Liverpool have fine university buildings, and the technical colleges of Wigan, Blackburn, and Burnley come but little behind the College of Technology

in Manchester. The venerable pile of buildings known as
Chetham's Hospital and Library near Manchester Exchange

Town Hall, Bolton

and Victoria Stations are educational buildings of another
kind. The ancient hall, the reading room, and the cloisters

mark it off at once from the modern educational buildings we have just mentioned.

Exchanges, etc.

The great Exchanges of Liverpool and Manchester are the very life-blood of South Lancashire. The Royal Exchange in Cross Street, Manchester, has a very grand facade, and recent extension and rebuilding has made this fine edifice much more visible than are others of Manchester's great buildings. Liverpool Exchange is built near the Town Hall, and with the latter building encloses the quadrangle known as "the Flags."

The Royal Liver Buildings, the Dock Board Offices, and the Custom House are noteworthy commercial buildings facing the Mersey at Liverpool, and add dignity to the river front.

14. Communications.

Roads, Canals, Railways.

We can only guess at any roads or tracks which existed in South Lancashire before the Roman period. We have already learnt that Manchester (Mancunium), Ribchester (Bremetenacum) and Colne (Colunio) were important centres during the Roman occupation, and that from them many roads radiated. The Romans were great road-makers, and the remains of their well-made roads may still be seen. On Blackstone Edge, east of Littleborough, the pavement is still in good condition. The stones in the middle of the road show very clearly the wearing effects of the brakes used on the steep gradients of the Pennine moorlands.

Royal Exchange, Manchester

After the Romans left Britain their roads were neglected and for long ages no new, good roads were made. All through the Middle Ages and into the early modern period, communications appear to have gone from bad to worse until, in the seventeenth century, there were few even moderate roads in South Lancashire and none that would now be called good. In many parts there were only ill-kept pack-horse tracks. We learn something about the roads in the seventeenth century from Agilby's *Britannia, a Geographical and Historical Description of the Roads of England and Wales,* published in 1675. This book was re-issued in a different form in 1699, and a new edition was edited by "Senex" in 1719. There are two main roads shown, the best being the London-Carlisle Road, and its branches. Some of the places on this main road were Warrington, Newton-le-Willows, Wigan, Standish, Leyland, Baumber Bridge; and branches are shown to Manchester, Liverpool, Holland, Chorley, and Blackburn. A second main road was the one from York to Chester, via Warrington and Manchester.

With the coming of the industrial era the need for better communications was soon felt. In the middle of the eighteenth century many new roads were begun and old ones improved. We read, for example, that in 1754 the road from Manchester to Rochdale and over the moors to Burnley was made into a "turnpike"; and similar changes are recorded in other parts of the county. That wonderful road-maker, John Metcalfe ("Blind Jack of Knaresborough"), planned several of the new roads in East Lancashire, and his work was done mainly between 1760 and 1790.

Arthur Young wrote his famous *A Six Months' Tour in the North of England* in 1770–71, and from it we may learn something about one of our "main roads." He wrote as follows in describing the road between Wigan and Preston. "I know not in the whole range of language terms sufficiently expressive to describe this infernal road,... let me most seriously caution all travellers who may accidentally propose to travel this terrible country, to avoid it as they would the devil; for a thousand to one they break their necks or their limbs by overthrows or breakings-down. They will here meet with rutts, which I actually measured, four feet deep and floating with mud, only from a wet summer: what, therefore, must it be after a winter?...I actually passed three carts broken down in those eighteen miles of execrable memory."

About this time, James Morris, "of Brock Forge, near Wiggan in Lancashire," had introduced broad wheels, up to 13 inches in width, so as to negotiate the muddy roads better; but what was really wanted was an improvement in the roads themselves.

We can see that many new roads had been made by the end of the century if we consult Cary's *New Itinerary of Great Roads*, the first edition of which was published in 1798. A revised, official edition was brought out in 1810 under the Surveyor and Superintendent of the Mail Coaches. This shows Manchester as the centre of a considerable system of roads, with many new ones leading into the hilly districts of East Lancashire. For example, there were now two fairly good roads from Manchester to Colne, one passing through Rochdale, Whitworth,

Bacup, Cliviger, and Burnley, the other through Bury, Walmersley, Crawshawbooth, Goodshaw Chapel, Burnley, Little Marsden, and Great Marsden. There was an important road from Preston to Blackburn, Padiham, Burnley, and Colne. It is interesting to note that the river which we now call the Darwen is called by its correct name, Derwent, on this map, and that the name of the town now known as Rishton is spelt "Rushton," which corresponds with the pronunciation still used by the older inhabitants of that district.

McAdam's new ideas were introduced into Lancashire between 1810 and 1820, and the influence of Telford and his pupils began to be felt immediately afterwards. Many of the "New Roads" (still known by that name) in Lancashire towns date from about 1820 to 1830.

So much for the roads. We must again go back to the early days of the industrial era for the beginnings of the extensive system of inland waterways which have had so much to do with South Lancashire's development. In the first half of the eighteenth century there were carried out the improvements known as the Mersey and Irwell Navigation, and the River Douglas Navigation, both of which received their Acts in 1720. The latter reduced the cost of carriage of coal from Wigan to Liverpool very considerably, although the distance travelled is very much greater. As a result of the former the cost of carriage of goods per ton between Manchester and Liverpool was reduced from 40s. to 10s. An Act was obtained in 1755 for deepening and straightening Sankey Brook, and so connecting Warrington and St Helens with the Mersey, and the canalised river was opened for traffic in 1760.

These early ventures paved the way for the more famous enterprise of Francis Egerton, third Duke of Bridgewater, who called in James Brindley, the illiterate engineer, to carry out his plans. The Duke had coal pits at Worsley, near Manchester, and realised that there was a growing demand for coal at the big town,—though it was a quarter of a century before the introduction of steam-engines into the Manchester district. In his Bill before Parliament the Duke promised that he would sell coal in Salford at a maximum price of 4*d*. per cwt., about half the price then current.

Brindley was original in his ideas and he superseded the Duke's plan of a system of descending and ascending locks across the Irwell at Barton by his famous aqueduct, which was afterwards regarded as one of the wonders of the North. The first boat-load was towed by horses on July 17th, 1761, and thousands of people came from far and near to see the unusual sight. An Act for the extension to Runcorn received the Royal assent in March, 1762. A portion of this longer canal was ready for traffic in 1767, and in 1772 a vessel of 50 tons went through from Runcorn to Salford; but the locks were not finally completed until 1776.

Meanwhile the Leeds and Liverpool Canal Act was passed in 1766. This canal connected Liverpool with Wigan, Chorley, Blackburn, Burnley, and Colne, and passed through the Craven Gap via Skipton to Keighley, Bingley, Shipley, and Leeds. There are several branches of this canal, one of the most important being that from Wigan, which connects it with an extension of the Bridge-water Canal at Leigh.

The Rochdale Canal was finally opened in 1804. This connects Manchester with Oldham, Middleton, Rochdale,

A MAP OF THE CANALS OF SOUTH LANCASHIRE AND DISTRICT, SHOWING THE CHIEF TOWNS AND CITIES WITH WHICH THEY ARE CONNECTED

and Littleborough, and proceeds over the Pennines at a height of 601 feet to Todmorden and Sowerby Bridge, where it joins the Calder and Hebble Navigation.

The Huddersfield Canal was a shorter one of $23\frac{1}{2}$ miles, made to link up Ashton and Huddersfield with the Calder and Hebble Navigation. It passed through the Pennines by the Standedge tunnel, $3\frac{1}{10}$ miles in length. Between Ashton and Manchester the canal joins the Peak Forest Canal and the Stockport Canal.

The Manchester and Bolton Canal was authorised in 1791. There was afterwards added a branch to Bury, and this canal therefore connects these important towns with the Bridgewater Canal at Manchester, and through that with all the other systems except the Sankey Canal, which has no through-connections.

But all the barge canals, useful and important though they have been, sink into insignificance by the side of the Manchester Ship Canal,—one of the greatest ventures of modern times, and one which has been of vast importance to the whole of the great industrial region of which Manchester is the centre. Manchester has now taken its place as about the fourth of the great ports of the British Isles, and, by means of the barge canals, it is in direct water communication with most of the industrial towns in South Lancashire, the West Riding of Yorkshire, and North Staffordshire, and has also through-connections with London, Bristol, Hull, and almost all parts of the Midlands of England.

The Ship Canal was opened for traffic in 1894, and the increase in traffic has been so rapid that, from a clearance of 926,000 tons in 1894, it had grown to 5,800,000 tons in 1913. Then came the Great War in 1914–18 and a serious decline, but in the year 1926 it had risen again to nearly 7,000,000 tons. The canal is $35\frac{1}{2}$ miles long, with its principal docks at Salford. It has a depth of 28 feet, and for nearly the whole of its length has a bottom width of

120 feet. Steamers of over 12,000 tons regularly navigate it. It joins the River Mersey at Eastham. The cost of making the canal was over £15,000,000. At Barton, the Bridgewater Canal is carried over the Ship Canal by an aqueduct, which is 6 feet deep, 18 feet wide, and has a span of 90 feet, in the middle of which there is a swinging

Swing Aqueduct and Road Bridge, Barton

span which has a weight of 1450 tons. The largest dock at the present time is the famous No. 9, which is 2700 feet by 250 feet in extent.

We may now turn to the railways, which were early established in South Lancashire. The famous Liverpool and Manchester Railway obtained its Act in 1826, and, though many other railways were already being discussed and the Stockton and Darlington Railway had actually been opened in 1825, yet the new venture across South Lancashire is regarded as the great pioneer railway. Construction was commenced in 1826, with George Stephenson as engineer. The engineering difficulties were great for those days, and, above all, a track had to be constructed

Steamship discharging in No. 9 dock, Manchester Docks

over four miles of water-logged and spongy bog-land. This "Chat Moss" difficulty was, however, overcome by Stephenson, who followed out a method used by "Blind Jack of Knaresborough" fifty years before. The famous tunnels under Liverpool also gave the engineers a great deal of trouble. The directors offered a prize of £500 for the best locomotive engine, which was to draw a load of at least three times its own weight, and travel with it at not less than ten miles per hour. In October, 1829, the trial took place at Rainhill, Liverpool, and four engines were entered. Stephenson, the engineer of the line, was there with his "Rocket," and this was the only engine to stand the test and satisfy the conditions. The prize was at once awarded to Stephenson, and in September, 1830, the railway was opened, a date which may well be called the beginning of the great railway era.

Between 1830 and 1850 there was a vast amount of railway construction in South Lancashire. As early as 1829 an Act was passed for the construction of the Kenyon and Leigh Junction Railway, the engineer being Rastrick. In the Goldsmiths' Library of the University of London may be seen a neatly written specification by Rastrick, a Civil Engineer, of Stourbridge, Worcestershire, in which is his autograph, "John U. Rastrick, C.E., at Liverpool, 16 Octr., 1829." Written in pencil on the fly-leaf by his son, is the following, "The first specification for a railway that was ever made."

In 1839 the Manchester and Leeds Railway was opened as far as Littleborough, and the summit tunnel (2956 yards in length) between Littleborough and Walsden was completed in 1840. The East Lancashire, the West Lancashire, the Manchester and Bolton, and the Liverpool and Bury Railways were in course of time combined with the Manchester and Leeds Railway, and the complex system of the Lancashire and Yorkshire Railway grew up.

The proprietors of the Manchester, Bolton, and Bury Canal, in considering the position of their undertaking in 1830, came to the conclusion that the day of canals was

over and, in 1831, they obtained an Act for converting
their waterway into a railway. Next year, however, they
repented and obtained an Act to construct a railway by
the side of the canal. The Manchester and Bolton Railway

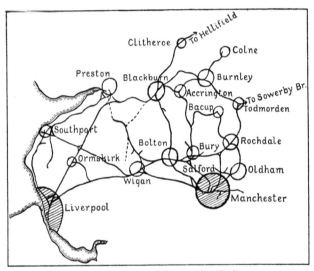

The former Lancashire and Yorkshire Railway system
so far as it served the northern part of South Lancashire

Railways shown by dotted lines were jointly owned

was opened for traffic in 1838, and the Bury branch was
abandoned for the time being. Bury had to wait for the
making of the Manchester, Bury, and Rossendale Railway,
which in 1845–47 grew into the East Lancashire Railway.

The Lancashire and Yorkshire Railway is now part of
"The London, Midland, and Scottish," which includes

all the railways of South Lancashire except the former Great Central and the Cheshire lines. For the sake of clearness, and because of the historical interest, the lines of the Lancashire and Yorkshire system are here given in

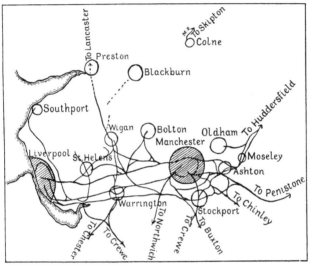

Railways of the southern part of South Lancashire, not part of the L.Y.R. system; they chiefly belonged to the L.N.W.R., M.R., G.C.R., and Cheshire Lines systems

one map, and the lines which belonged to the other systems in a separate map.

15. The Roll of Honour.

In considering some of the more famous men and women who have been connected with South Lancashire it has been thought fit, in view of its being the most important industrial region, to take some of her great inventors first.

One of the earliest of South Lancashire's great inventors was John Kay, of Walmersley, near Bury. Born in 1704, he was a reed-maker in Bury in 1730, when he took out a patent for a twisting and carding machine, and about the same time made reeds of metal instead of cane. His great invention however (1733) is that of the "fly shuttle," which was driven across the shed of warp by means of a mechanical "picker." It revolutionised the art of weaving, and had the usual reception given to labour-saving devices, for his house was wrecked by a mob in 1753. His son, Robert Kay, followed his father's bent and produced the "drop box" for shuttles about 1759, an invention which made it possible to weave with weft of different colours without stopping the loom to change the shuttles. James Hargreaves, a working carpenter and hand-loom weaver, was born about 1745 at Stanhill, near Blackburn. In the year 1760, he was employed by Robert Peel (grandfather of the great statesman) to make an improved carding-machine. He made his new spinning machine, which he named after his wife the "Spinning Jenny," in 1764–66. By it the rapidity of spinning was at once increased many-fold, and he accordingly suffered the fate of Kay, for a Black-burn mob wrecked his machines in 1768. He removed to Nottingham and died in 1778.

Meanwhile, Richard Arkwright, born in Preston in 1732, settled in Bolton, as a barber, some 18 years later. Here he showed an inventive turn of mind, and enlisting the help of a clock-maker named Kay, began experiments on a spinning machine which seemed so promising that he and Kay returned to Preston. He erected his first spinning mill at Cromford in Derbyshire in 1771, where he could get plenty of water-power to drive his machines. In 1775

Sir Richard Arkwright

he made a machine for carding, drawing, roving, and spinning, all in one set; and in 1790 he installed one of Boulton and Watts' steam engines in a mill at Nottingham. He, too, suffered the usual fate, his works being destroyed by the mob, but he triumphed over all difficulties, was knighted in 1786, and died in 1792 worth half a million.

Samuel Crompton, who was born at Firwood ("Hall-i'-th'-Wood"), Bolton, 1753, had to use the Spinning Jenny in his regular work, and from about his 22nd to his 27th years he laboured to improve it. He completed his machine,

Samuel Crompton, inventor of the Mule Spinning Frame

which could spin very fine yarn, in 1779, and called it the "mule," because it was a combination of the wheel of Hargreaves, and the roller of Arkwright. The "carrier" was his own peculiar contribution. But his invention was stolen, he met with many misfortunes, and died in 1827 in comparatively poor circumstances.

The Peels were associated with the cotton trade of East Lancashire from a very early date. The first famous Robert Peel was known as "Parsley Peel" in the cotton region of Lancashire, because of the well-known parsley-leaf pattern which he printed on calico. The third son of this Robert Peel was also named Robert, and he and his partner, Yates, owned one of the most famous mills in Lancashire and revolutionised the cotton-printing industry. The printed calico industry became firmly established in East Lancashire and many other famous men were associated with it. One of these was John Mercer, born at Dean, near Great Harwood, in 1791. A weaver, and afterwards a calico-printer, he became so successful that in 1818 he was engaged as chemist in the colour shop of his old works, and in 1825 taken into partnership. He introduced many new processes into calico-printing; but the invention for which he is best remembered is that of "mercerising" cotton, which he brought out in 1848. Incidentally, the same experiments which led to this discovery showed him how to make "parchment paper," which he patented in 1850.

One of the greatest of the early chemists was John Dalton, who was born at Eaglesfield, near Cockermouth, in 1766. Originally a stocking weaver, he explored almost every realm of science. In 1793 he went to Manchester, where the rest of his life was spent in teaching and in scientific discovery. His famous papers, which have had so great an influence on modern chemistry, were read before the Manchester Literary and Philosophical Society in 1801 and 1803. He was the first to describe colour-blindness. In 1808 there was published at Manchester his great work, *A New System of Chemical Philosophy*, in which he explained his well-known Atomic Theory, by which he is best known. He died in 1844. Before the same Society there were read, in 1840 and 1843, the epoch-making

John Dalton's birthplace

papers of James Prescott Joule, who became as well-known in Physics as Dalton had been in Chemistry. Joule was born in Salford in 1818, and studied with Dalton in the later years of the chemist.

One of the earliest of famous engineers, James Brindley (born 1716), was not Lancashire born though most of his work was in that county. Beginning as a millwright he was called in by Francis Egerton, the third Duke of Bridge-water, to help in the making of the famous canal in 1759. The most daring of his plans were the great aqueduct over the Irwell at Barton, and the canal tunnels into the low hills at Worsley. After the success of the short canal, he undertook the longer Bridgewater Canal from Manchester to Runcorn. He became the most famous canal engineer of his time and his services were in constant demand.

George Stephenson has been dealt with in the chapter on Communications. He was born near Newcastle, in 1781, and his work was carried out in different parts of the country, but there is especial interest in the engineering of the Liverpool and Manchester Railway, and in the success of his engine "The Rocket." He was also chief engineer for the Manchester and Leeds Railway.

We come now to those who have won literary distinction. Thomas de Quincey was born at Greenheys, near Manchester, in 1785, and went to Manchester Grammar School in 1801. He entered later at Worcester College, Oxford, but left without taking a degree. At Grasmere he became acquainted with Wordsworth and Coleridge. He was a prolific writer, and is perhaps best remembered by his *Confessions of an English Opium-Eater*. He died in 1859. Another pupil of Manchester Grammar School who achieved literary distinction was the novelist William Harrison Ainsworth, born in King Street, Manchester, in 1805. His best novels, apart from *Old St Paul's*, are generally considered to be those with a peculiarly Lancashire flavour. His *Lancashire Witches* is one of these that has been widely read.

Mrs Gaskell, though born in London, lived a good deal at Knutsford (Cheshire), the "Cranford" of her famous novel, and lived also in Manchester. Mrs Hemans (1793–1835), famous for her lyrics, wrote, among other poetry,

Thomas de Quincey

Hymns for Childhood, and *Scenes and Hymns of Life*. She was born in Liverpool, but lived much in North Wales and died in Dublin. Philip Gilbert Hamerton, the essayist and art critic, was born at Shaw, near Oldham, in 1834.

The humour and peculiarly expressive dialect of the dwellers in the eastern part of South Lancashire have found expression in the work of many local writers. John Collier ("Tim Bobbin") was born at Urmston, near Manchester, in 1708. At first a weaver's apprentice he finally became Headmaster of the Free School at Milnrow. He carefully studied the homely Lancashire speech, and wrote historical and other sketches in that dialect. Another notable writer of Lancashire dialect, Edwin Waugh, was born at Rochdale in 1817.

Among earlier writers should be mentioned John Byrom, who was born at Kersal in 1691 or 1692, and eventually became Fellow of Trinity College, Cambridge. He invented a system of shorthand, and was elected a Fellow of the Royal Society, but he is best known to-day by his hymn, "Christians, awake, salute the happy morn."

A number of Lancashire men have made themselves famous as writers of history, generally of their own county or district, but in one case at least attaining to national distinction. Christopher Towneley was of the famous family associated with Towneley Hall, near Burnley, and was born in 1604. He made extensive researches into the history of Lancashire, and the Towneley MSS. left by him have long been a useful mine of information for students of Lancashire life. A later member of the same family was Charles Towneley, who was born at Burnley in 1737. He was a great collector of antiquities, and especially of Greek and Roman sculpture. The main part of his collection was bought for the British Museum in which the "Towneley Marbles" form a most important section.

John Whitaker, the son of a Manchester innkeeper, was born in 1735. After spending eight years in the Grammar School of his native town he won a school Exhibition at Brasenose College, Oxford. Later he transferred to Corpus Christi, of which college he afterwards became a Fellow. He wrote a *History of Manchester* in 1771, which was revised and enlarged in 1773 and again in 1775. His history was very much criticised, even in his own day, but it has served as a starting-point for much subsequent writing. Another Whitaker (apparently no relation) was the famous Thomas Dunham Whitaker, of the old family of Whitaker, of Holme, near Burnley. He was born in 1759, and after being in the hands of private tutors at Shaw and Grassington, he entered St John's College, Cambridge, in 1775, and took Orders. His local histories of Whalley, Craven, and Richmondshire are well known, and invaluable for the regions with which they deal.

Sir James Allanson Picton was born at Liverpool in 1805. He was an authority on land arbitration, and was architect of many important buildings in Liverpool. He took a great interest in educational and philanthropic movements, and worked hard and successfully for the establishment of a public library and museum, and was the author of the *Memorials of Liverpool*. The "Picton Reading Room," so-named by the Corporation of Liverpool in his honour, is a testimony to the great esteem in which he was held by his fellow-citizens.

Notable among divines was John Tillotson, who was a pupil at Colne Grammar School about 1640, and who took his degree at Cambridge about 1650. He was a great preacher, and had a great belief in the power of the pulpit. His sermons were translated into other languages and had a very large circulation. He became Archbishop of Canterbury in 1691.

Lancashire has produced a number of men who have made their mark in the service of their country, or in furthering what they considered to be its best interests. William Roscoe was born in Mount Pleasant, Liverpool, in 1753. He was a determined worker for the abolition of the slave-trade, and when he was elected Member of Parliament for Liverpool in 1806 he threw all his energy into that movement. He did not forget his native town, and the Liverpool Botanic Gardens and the famous Athenaeum Library are his best monuments in Liverpool.

One of Liverpool's most famous men, born in Rodney Street, almost within a stone's throw of Roscoe's birthplace, was William Ewart Gladstone. He was born in 1809, a son of Sir John Gladstone, a Liverpool merchant. Young Gladstone went to Eton in 1821, and to Christ Church, Oxford, in 1829. He took a double-first in Classics and Mathematics. He entered Parliament at the early age of 23 as a Conservative, and for nearly sixty years was a Member of the House of Commons. He gradually turned towards Liberalism and became probably the most famous of all Liberal statesmen, being many years Prime Minister. He was a fine orator, a great writer, and a successful administrator. Perhaps his greatest work was as Chancellor of the Exchequer, in which capacity he freed from taxation many articles of common consumption.

Two great statesmen associated with South Lancashire were John Bright and Richard Cobden, best known in connection with Free Trade. John Bright was born at Rochdale, in 1811, and was educated chiefly at schools under the control of the Society of Friends. He entered his father's business—the woollen trade—and took an active part in Rochdale's affairs. He joined Richard Cobden in the agitation for the repeal of the Corn Laws in 1839, and became the great orator of the cause. Bright's great co-worker in the Corn Law agitation was Richard Cobden, the son of a Sussex yeoman farmer. He was born at Midhurst, in 1804, and went to a Yorkshire school. Cobden became associated with Lancashire's great industry, and with the problems of a great industrial district. In 1831 he and his partners leased an old mill at Sabden and began the production of printed calicoes. He is chiefly known by the successful campaign in 1846 for the repeal of the Corn Laws.

The mention of the agitation for the repeal of the Corn Laws leads us to another great Prime Minister who was born in Lancashire. We have mentioned the Peels already in connection with the cotton trade. The second Robert Peel was the first Baronet, and his son was the famous Prime Minister. This third Robert Peel was born near Bury, in 1788, and went to Harrow, and then to Christ Church, Oxford, where like Gladstone he took a double-first, and entered Parliament at an even earlier age than Gladstone. He had a great career in Parliament and his name will be always associated with Catholic emancipation, the organisation of the Police, banking reform, and the repeal of the Corn Laws. He was the most prominent figure in British politics for a long time.

There have been many Stanleys, members of the great house whose ancestral home is Knowsley near Liverpool, who have played an important part in the affairs of their country. One of these has already been mentioned in connection with the Civil War of the 17th century—the Earl of Derby who was executed at Bolton as a reprisal for the share he had taken in the "Bolton Massacre." A much later member of that historic house was the nineteenth century statesman who was born at Knowsley, in 1799. He was educated at Eton, and at Christ Church, Oxford, and entered Parliament when he was twenty-two. He was a fine orator, and though he opposed the Liverpool and Manchester Railway, he brought in the measure for the freeing of the West Indian slaves in 1833, and was three times Prime Minister, in 1852, 1858, and 1866.

Another Lancastrian who deserves mention is Francis Towneley, the famous Jacobite. Born in 1709, he resided much in France, and his sympathies were with the exiled Stuarts. He joined the rising of 1745, and raised the "Manchester Regiment." When Prince Charles entered Manchester, he was accompanied by Towneley, who also went with him to Derby. At the collapse of the rising, Towneley was one of those who were taken; he was tried in July, 1746, was executed, and his head was placed on a pike at Temple Bar.

Some names remain to be mentioned as special benefactors.

One of the earliest of these was Hugh Oldham. After a successful University career he was, by turns, clergyman and schoolmaster, and finally, in 1505, Bishop of Exeter. He was a co-founder of Corpus Christi College, Oxford, and himself founded Manchester Grammar School in 1515. The early history of the school was a chequered one, but the good Bishop builded better than he knew, and who would be bold enough to estimate what the school has been, is, and will be, to South Lancashire?

A century later than Hugh Oldham was Humphrey Chetham, born at Crumpsall, in 1580, and educated at the Grammar School. He became a Manchester merchant, and he and his brother, having amassed much wealth, bought land in and about the city. He did great things for his county during his life-time, and when he died he left £1000 for the foundation and endowment of a hospital for the education and maintenance of 40 poor boys, and also left £1000 for a Library. The Hospital School and Library are among the most famous and most useful of Manchester's institutions.

Another South Lancashire man who remembered his county was William Hulme, who was born at Kearsley, in 1630. He was a scholar in Manchester Grammar School, and afterwards a student at Brasenose College, Oxford. By his will, he devised a trust estate consisting of property in Manchester and district. The property left under this trust has become exceedingly valuable and various Acts of Parliament and schemes of the Charity Commissioners have dealt with the enormous income which is now available for disbursement by the Hulme Trustees.

John Owens was a wealthy Manchester merchant, who, in 1846, left nearly £100,000 for the establishment of a

University College, which should be free from religious tests. The Owens College was founded in 1851, and out of it has grown the Victoria University of Manchester.

We must not overlook the famous banking family of the Heywoods, of Manchester, who have done so much for the education of that city. The father, Sir Benjamin Heywood, and the sons, James Heywood and Oliver Heywood, were all active and continuous in their support of popular education.

The great sister and rival city, Liverpool, has also had its friends and benefactors. The two William Rathbones, father and son, have done for Liverpool what the Heywoods did for Manchester. William Rathbone, the elder, was born in 1757, and was a member of the famous Liverpool firm of William Rathbone and Son. To this firm came the first consignment of raw cotton ever shipped to Liverpool from the United States. His son bore the same name and carried on his father's traditions, with the additional advantage that he had been at the University of Oxford. He was a staunch advocate of Catholic emancipation, of municipal reform, and of popular education.

The last name which we shall mention is that of Sir Joseph Whitworth, who was born at Stockport, in 1803. He founded the great engineering works at Openshaw, Manchester, and amassed a great fortune. Whitworth's name might well have come among the inventors, but he is probably remembered most by the famous Whitworth Scholarships, for the endowment of which he left £100,000. The Whitworth Technical Institute, at Manchester, has now become the Manchester School of Technology.

16. The Chief Towns and Villages of South Lancashire.

(The figures in brackets give the population in 1921. C. = City.
C.B. = County Borough. N.C.B. = Non-County Borough.
U.D. = Urban District. The figures at the end of each section
refer to pages in the text.)

Abram, U.D. (6858), lies immediately south of Wigan. The industries are those of Wigan.

Accrington, N.C.B. (43,610), is situated just north of the Rossendale Fells. It is mainly a modern well-built industrial town. The industries are cotton-spinning and weaving, calico-printing, paper manufacture, slipper-making, coal-mining, the making of bricks, tiles, and sanitary ware, chemical manufactures, and engineering. The Church of St James, formerly a chapel-of-ease of Whalley, was rebuilt in 1763. The fine Town Hall was erected as a memorial to Sir Robert Peel. (pp. 16, 21, 24, 25, 69, 71, 72, 76, 79, 82, 85, 105, 110.)

Adlington, U.D. (4393), lies south of Chorley. The chief industries are cotton-spinning, cotton-weaving, calico-printing, and bleaching and dyeing. (p. 25.)

Ashton-in-Makerfield, U.D. (22,489), is about 12 miles west of Manchester and 3½ miles south of Wigan. There is coal-mining, and the manufacture of nails, files and locks. A Roman Road to the north passed through the district, and traces of it have been found. Some relics of the Roman occupation are in the museum at Knowsley Hall. (p. 80.)

Ashton-under-Lyne, N.C.B. (43,333), is about six miles east of Manchester, in the south-eastern corner of the county. The industries are cotton-spinning, cotton-weaving, dyeing, silk-weaving, hat-making, and various branches of engineering. The church contains some fine medieval stained glass. (pp. 6, 26, 27, 78, 79, 81, 85, 95, 105, 106, 111.)

Aspull, U.D. (7851), is 3½ miles north-east of Wigan. The industries are coal-mining, and cotton-spinning.

Atherton, U.D. (19,863), lies about 10 miles west-north-west of Manchester, and 4½ miles south-west of Bolton. It has collieries, iron-works (where nails, bolts, and spindles are made), cotton-spinning and cotton-weaving. In the Civil Wars of the seventeenth century two small battles were fought at Chowbent, a village near Atherton. (p. 80.)

Audenshaw, U.D. (7878), lies immediately to the south-west of Ashton-under-Lyne. Its industries are cotton-spinning, cotton-weaving, finishing of cotton goods, hat-making, and engineering.

Bacup, N.C.B. (21,256), is 7 miles north of Rochdale, and 6 miles south of Burnley, in the heart of the Rossendale Fells. The chief industries are cotton-weaving, cotton-spinning, the making of felts, slippers, and light shoes, and stone quarrying. On the north side of the town are the remains of an ancient encampment, known as Broadclough Dyke. (pp. 16, 26, 27, 38, 51, 71, 77, 85, 88, 103, 110.)

Barrowford, U.D. (5624), is situated 1 mile north of Nelson, on Pendle Water. Its industries are cotton-weaving, and cotton-dyeing. (pp. 24, 85.)

Billinge, U.D. (5162), is about 5 miles south-west of Wigan. It is a coal-mining district, in the middle of good agricultural land, producing oats, root-crops, and potatoes.

Blackburn, C.B. (126,630), lies at the north-western edge of the Rossendale Fells, about 24 miles from Manchester. Six branches of the L.M.S.R. converge on Blackburn, and the Leeds and Liverpool canal passes through the town. It is essentially a weaving town, and specialises in cotton goods for India and China. There are also large engineering works and flour-mills. The town gave the name to the Blackburn Hundred, one of the divisions of Lancashire. At one time the hundred was known as Blackburnshire. There are many fine modern buildings among which the Town Hall, the Exchange, the Technical College, the Library, Museum, and Art Gallery, the Market Hall, and the new Public Hall may be mentioned. Hargreaves, the inventor of the Spinning Jenny, was a

St Mary's Parish Church, Blackburn

Blackburn man, and set up his first machines there. The first Sir Robert Peel was born at Blackburn, but afterwards removed to the Irwell Valley. (pp. 3, 5, 7, 19, 21, 24, 25, 38, 58, 63, 68, 79, 82, 83, 85, 91, 93, 95, 97, 101, 103–105, 110–112.)

Bolton, C.B. (178,678), is one of the largest, most prosperous, and best built of the Lancashire industrial towns. It lies, as its full name Bolton-le-Moors indicates, among the Rossendale Fells, about 11 miles north-west of Manchester. The railways include six branches of the L.M.S. The Bury and Bolton Canal also serves the town. There is a great deal of cotton-spinning (especially of fine yarns), and some weaving. There are large engineering works, tanneries, paper-mills, and chemical works; and coal-mining is carried on in the outskirts of the town.

Market Cross, Bolton

Bolton played a considerable part in the Civil Wars of the seventeenth century, and the Earl of Derby was executed in the streets of the town in 1651. Two great inventions in the cotton trade have

Hall-i'-th'-Wood, Bolton

(from the painting by William Linton)

been associated with Bolton. In the years 1766–68 Richard Arkwright worked out his "water-frame" for spinning cotton, and Crompton, who lived at Hall-i'-th'-Wood, a fine specimen of old timbered architecture, during the years 1774–79, invented the mule-frame for spinning fine cotton yarns. His house is now the property of the town and has been made into a museum, not only of Crompton's work, but of the furniture and fittings of the last two hundred years. Among modern buildings is the handsome Town Hall completed in 1873. (pp. 5, 7, 19, 21, 26, 27, 38, 51, 58, 65, 68, 71, 72, 76, 77, 79, 81, 85, 86, 88, 92, 93, 97, 98, 105, 106, 109–112, 114, 123.)

Bootle, C.B. (76,508), is situated to the north of Liverpool, and though a separate County Borough, is really a continuation of that city. It has some of the great Mersey docks, and there is a very considerable industry in various forms of marine engineering, and in repairs of engines and ships' fittings. The borough possesses a fine Town Hall, a Technical School, and the large and beautiful Derby Park. (pp. 5, 80, 97.)

Burnley, C.B. (103,175), is about 22 miles due north of Manchester, and lies at the north-eastern corner of the Rossendale Fells. It is especially a cotton-weaving centre and claims to produce more yards of cloth than any other town in the world. Its speciality is the cheap, plain cotton cloth, the woven pieces of which are known as "Burnley Lumps." There are about 87,500 looms in the town and in recent years the manufacture of flannelettes and other more expensive goods has increased considerably. There is some cotton-spinning, cotton-dyeing and bleaching, and paper manufacture. Numerous engineering works make looms and other textile machinery chiefly. There is a good deal of coal-mining, and coke-making for iron-works. Bricks and tiles and sanitary ware are also made.

The parish church of St Peter has a fifteenth-century tower, and the grammar school close by was founded in the sixteenth century.

9-2

Close to the school is a cross known locally as the Cross of Paulinus. Among its modern buildings is a fine Technical College.

In the valleys around Burnley are many interesting old houses.

Municipal College, Burnley

Holme, once the residence of Dr Whitaker, Master of St John's College, Cambridge, and of Dr T. D. Whitaker the historian; Towneley Hall, now a municipal Art Gallery and Museum; and Gawthorpe Hall, the residence of Lord Shuttleworth. It is said that the poet Spenser wrote his *Shepherd's Calendar* at Hurstwood, and a house there is called Spenser's House. (pp. 5, 7, 13, 14, 16, 19–21,

23–25, 38, 51, 58–61, 63, 67, 71, 72, 76, 77, 79, 80, 82, 83, 85, 86, 88, 91–93, 97, 101, 103–105, 110, 119, 120.)

Burscough (Lathom and Burscough, U.D., 7663) is 3½ miles north-east of Ormskirk. The country around is flat, agricultural land. Near by are the remains of an Augustinian Priory founded in the twelfth century. (pp. 63, 90, 95.)

Bury, C.B. (56,426), stands on slightly raised ground between the Irwell and the Roche about 10 miles north of Manchester. The industries include cotton-spinning and weaving, bleaching, dyeing and printing, felt-making, paper manufacture, chemical manufacture, and engineering. The handsome Drill Hall accommodates over 7000 people. Sir Robert Peel, the famous Prime Minister, was born at Chamber Hall near here. (pp. 5, 7, 19, 26, 27, 38, 58, 68, 76–80, 85, 86, 88, 89, 93, 95, 97, 103, 105, 106, 109, 110, 112, 122.)

Chorley, N.C.B. (30,576), is almost equidistant from Preston, Blackburn, Bolton, and Wigan, on the Lancashire plain, close to the western edge of the Rossendale Fells. It is an old market town which in the last hundred years has developed into a modern manufacturing town, its principal industries being cotton-weaving, cotton-spinning, cotton-bleaching and printing, chemical manufactures, and engineering. In the neighbourhood are two famous old houses, Astley Hall and Duxbury Hall. It was probably from the latter that Myles Standish went to the New World in 1620. (pp. 25, 71, 76, 77, 82, 85, 101, 104, 105.)

Clitheroe, N.C.B. (12,204), lies in Middle Ribblesdale, to the south of the river. Its chief industries are cotton-spinning, cotton-weaving, calico-printing, and limestone quarrying. It has the remains of a Norman castle built on the top of one of the limestone knolls for which the district is famous. This was erected by a De Lacy in the eleventh century, and was dismantled in 1649. A mile north of the castle is Brungerley Bridge where the fugitive king, Henry VI, was captured in 1464—"byside Bungerley hipping-stones in Lancastreshyre." (pp. 6, 10, 13, 14, 20, 38, 51, 55, 59, 63–66, 71, 82, 85, 89, 110.)

Colne, N.C.B. (24,755), is the most easterly of the "cotton towns" of north-east Lancashire. The industries are cotton-weaving and spinning, dyeing, tanning, and engineering. The name Colne

Pack Horse Bridge, Wycoller

is usually held to be the Latin *Colunio* or *Colonia*, and Colne was certainly a Roman station. On a hill near, now called Caster Cliff, is an oval earthwork which has been described as a Roman encampment and as a prehistoric fortress. The parish church of St Bartholomew dates from the twelfth century, but there is little of the original structure remaining. Near the church is the building known as the Colne Grammar School. This has replaced an earlier building in which Archbishop Tillotson received some part of his education. Not far away is Wycoller Hall, famous for its association with Charlotte Brontë's *Jane Eyre*. (pp. 5, 6, 13, 15, 16, 20, 24, 38, 44, 54, 55, 57–59, 63 65, 71, 77, 80, 82, 83, 85, 89, 91, 95, 99, 102–105, 110, 111, 120.)

Crosby (Great Crosby, 13,722), which is mentioned in Domesday Book, is now a residential suburb of Liverpool, and is 6 miles

Wycoller Hall, near Colne

northward of the centre of the city. Crosby Hall, the residence of the Blundell family, is at Little Crosby, and there is another famous hall at Ince Blundell a little farther north. The Merchant Taylors' School at Great Crosby was founded in 1619, out of money left by John Harrison, a merchant tailor of London, whose father was a native of the village. (p. 55.)

Darwen (Over), N.C.B. (37,913), is 4 miles south of Blackburn and 8 miles north of Bolton. It lies near the north-western edge of the Rossendale Fells, and is drained by the River Darwen (= Derwent).

Cotton-spinning and cotton-weaving are the most important industries, but there are paper-mills and paint manufactures, as well as minor chemical industries. Darwen has one of the few sets of iron furnaces of South Lancashire, and brick and tile making is also carried on in the district. (pp. 21, 24, 25, 55, 71, 76, 77, 82, 85.)

Denton, U.D. (17,631), is about 5 miles south-eastward of Manchester. Its chief industry is the making of men's hats. There is also coal-mining in the neighbourhood. (p. 88.)

Downham, 3 miles north-east of Clitheroe, is a pretty village lying among limestone knolls at the foot of Pendle. The Roman road from Ribchester to Ilkley passes through it. Downham Hall is the seat of the Asshetons, one of whom wrote an interesting diary on the life of a country gentleman in these parts in the reign of James I. (pp. 23, 39, 46, 47, 59.)

Droylsden, U.D. (13,877), is practically a suburb of Manchester. The industries are cotton-spinning, dyeing and finishing, chemical manufactures, engineering, and brick-making. (pp. 26, 71.)

Earlestown, U.D. (10,077), is 5 miles north of Warrington and 4 miles east of St Helens. The chief industries are the making of railway-wagons, and glass manufacture, sugar-refining, and coal-mining.

Eccles (*ecclesia* = a church), N.C.B. (44,237), lies immediately to the west of Salford, on the Manchester Ship Canal and the Bridgewater Canal. The industries are cotton-spinning and cotton-weaving, rubber works, engineering, and the famous Nasmyth Steam Hammer Works. There are two famous engineering triumphs at Eccles: the first is the aqueduct by which Brindley carried his canal across the Irwell; the second, the contrivance for carrying the Bridgewater Canal across the Ship Canal. A box-like structure full of water and weighing 1600 tons is made to revolve on a cylindrical pillar and so the Ship Canal is cleared for shipping. The parish church was founded in the twelfth century and still contains many features of interest. (pp. 63, 80, 81, 95.)

Failsworth, U.D. (16,972), lies between Manchester and Oldham. Its chief industries are cotton-spinning, silk-weaving, and hat manufacture. (p. 26.)

Farnworth, U.D. (27,901), lies 3 miles east of Bolton. The industries are cotton-spinning, cotton-weaving, bleaching, dyeing and finishing, coal-mining, and engineering. Great Lever Hall, the ancient seat of the Asshetons of Middleton, was mentioned by Pepys in his *Diary* (1662). (pp. 27, 85.)

Formby, U.D. (6319), is a residential outlier of Liverpool about 12 miles to the north-north-west of the latter. The original Formby (Formebie in Domesday Book) was a fishing village, which was finally destroyed by the advancing sand in the eighteenth century. Formby Hall, a fourteenth or fifteenth century house, is about 2 miles north of the modern village. (pp. 7, 32, 35, 51, 55, 93.)

Great Harwood, U.D. (13,596), is about 4 miles north-east of Blackburn. Its chief industry is cotton-weaving. In the neighbourhood was born John Mercer who invented the process of "mercerising" cotton. The church has several features of interest, notably its fifteenth-century tower; and its register is practically complete from 1579. (pp. 24, 69, 85, 115.)

Haslingden, N.C.B. (17,485). This town lies well up on the Rossendale Fells, much of it being above the 600 feet contour line. Its industries are cotton-weaving, and especially the cotton-waste trade. There are engineering and chemical works, dye and bleach works, and paper-mills in the vicinity. (pp. 16, 26, 27, 71, 76, 85.)

Haydock, U.D. (10,333), lies about 6 miles north-north-west of Warrington and 4 miles east of St Helens. It is a coal-mining town in the middle of a farming district which produces potatoes, root-crops, oats, and clover.

Heywood, N.C.B. (26,691), lies midway between Bury and Rochdale near the southern edge of the Rossendale Fells. It has cotton-spinning, cotton-weaving, paper-mills, and chemical works.

There are also boiler works and railway-wagon works. (pp. 27, 76, 77, 81, 85, 105.)

Hindley, U.D. (23,574), is about 2½ miles east of Wigan. Its industries are coal-mining, cotton-spinning, and engineering.

Horwich, U.D. (15,616), lies at the western edge of the Rossendale Fells. The locomotive works of the old L.Y.R. are here. Not far away is the prettily situated Rivington and Blackrod grammar school, almost under the shadow of Rivington Pike. There is a beautiful park between the steep western edge of the hills and the large reservoirs belonging to the city of Liverpool. (pp. 21, 80.)

Hulton (Little Hulton), U.D. (7911), lies to the southward of Bolton. Its industries are coal-mining and cotton-spinning.

Ince-in-Makerfield, U.D. (22,865), lies to the south-east of Wigan, of which it practically forms a part, and in whose industries it shares.

Irlam, U.D. (9471), is adjacent to the Ship Canal about 8 miles west-south-west of Manchester. The industries are chemical manufactures, soap and candle works, engineering, and cotton-spinning. (p. 76.)

Kearsley, U.D. (9610), has the industries of Farnworth and Bolton, viz. coal-mining, cotton-spinning, doubling, and weaving, bleaching, dyeing and finishing.

Lees, U.D. (4789), is south-east of and contiguous to Oldham. Its industries are those of Oldham, of which it is practically a suburb.

Leigh, N.C.B. (45,545), is about 11 miles west of Manchester. A branch of the Bridgewater Canal meets a branch of the Leeds and Liverpool Canal here. The industries are coal-mining, engineering, cotton-spinning and cotton-weaving. (pp. 81, 85, 105, 109.)

Lever (Little Lever), U.D. (5194), is 3 miles south-east of Bolton. It has coal-mines, chemical works, paper-mills, and cotton-spinning.

Leyland, U.D. (9034), is about 5 miles south of Preston. It is an old English village which is rapidly becoming a modern manu-facturing town. The industries are calico-printing, chemical works, rubber works, and the making of motor-wagons. (pp. 3, 25, 63, 78, 79, 95, 101.)

Litherland, U.D. (16,383), lies immediately to the north of Bootle, of which it may almost be regarded as a suburb.

Littleborough, U.D. (11,488), is at the southern edge of the Rossendale Fells, about 3 miles north-east of Rochdale. The manufactures are cottons, woollens, chemicals, rubber, and asbestos. The Summit Tunnel of the old L.Y.R. is a little to the north of the town. Blackstone Edge, with its Roman road, is near. (pp. 20, 27, 57, 59, 71, 78, 85, 99, 105, 109.)

Liverpool, C. (803,118), is the second port, and the third city in population, in the United Kingdom. It is situated on the Mersey estuary opposite the narrow neck, and its unequalled series of docks stretches for about 8 miles from north to south (the city facing west). It is by far the greatest cotton-importing place in the world; and there are also vast imports of timber, wood-pulp, grain, flour, and prepared foods; bacon and ham, butter, eggs, and live stock; fresh fruits and tinned fruits; raw and refined sugar, molasses; tobacco; vegetable and animal fats and oils. The manufactures arise mainly from the imports, and include sugar-refining, chocolate manufacture, flour-milling, biscuit manufacture, the making of jam and marmalade, oil-seed crushing, and soap-making. There are also engineering works, clothing factories, works where furniture is made, and a host of minor trades.

Liverpool has a flourishing University; the city is the seat of a Bishopric of the Church of England, and of an Archbishopric of the Roman Catholic Church. The name goes back to 1173, and

King John granted a charter in 1207. A second charter was granted by Henry III in 1229, and this recognised the Guilds of the time with their privileges. In the latter half of the thirteenth century a castle was built which fell into ruin in the sixteenth century and seems to have been finally demolished in 1720. Leland, during his official journeys of inspection from 1534 to 1540, came to

Castle Street, Liverpool

Liverpool, and in his *Itinerary* he speaks of "Lyrpole alias Lyverpoole," and he mentions the "Castelet" and the stone house of the Earl of Derby. He also tells us that much Irish yarn was bought there by Manchester merchants. Liverpool had its full share of the decline of fortune which followed on the dissolution of the monasteries. It was not until the later part of the seventeenth century that real prosperity began, and from that time it went ahead at a great pace. Irish and American trade, and increased

Liverpool Cathedral

manufactures in Lancashire, all contributed to this rise in prosperity. The population at the end of the seventeenth century was about 5000, and the port was described as thriving and prosperous. In 1709 there was begun the first "floating dock," and modern Liverpool may be said to date from that time. It was now the third port in the kingdom. During and after the great Napoleonic Wars it grew in importance even more rapidly, and when steamships entered the Mersey, and a railway was made to connect it with Manchester, it rapidly climbed to the second place among ports.

Liverpool has few relics of medieval times; modern "progress" has cleared them away. It possesses, however, some of the finest modern buildings in Europe, among which the following may be mentioned: St George's Hall, the Walker Art Gallery, the Picton Reading Room, the Free Library, the Technical School and Museum, the Cotton Exchange, the Town Hall (1754), the University, the Philharmonic Hall, the Royal Liver Buildings, and the Post Office. The Cathedral, when completed, will be the largest in the country. (pp. 5–7, 16, 22, 28, 29, 32, 33, 37, 38, 55, 56, 58, 65, 67, 70, 73, 75, 77, 78, 80, 88, 89, 93, 95, 97, 99, 101, 103–105, 109–111, 117, 118, 120, 121, 123, 125.)

Manchester, C. (730,551, and see Salford). The Geological Map of South Lancashire shows a kind of bay of New Red Sandstone with Coal Measures to the north and east. In this bay stands Manchester. The topographical map shows a number of river valleys converging towards Manchester; the central position of the city is thus emphasised and its importance explained.

The city stands at the junction of the Irk and the Medlock with the Irwell. It is a great railway centre and has four termini of the first rank. The canals are the Manchester Ship Canal, the Bridgewater Canal, the Bury and Bolton Canal, the Rochdale Canal, the Manchester and Ashton Canal connecting to the Huddersfield Canal, the Peak Forest Canal, the Macclesfield Canal, and the Stockport Canal.

Manchester is the focus of the vast cotton industry. It is the market, the banking centre, the warehousing and the distributing centre. On Tuesdays and Fridays, the "Market Days," over 10,000 manufacturers and merchants come to Manchester to buy and sell cotton goods and things connected with cotton manufacture. There are also some manufactures carried on, especially in the suburbs. Spinning, weaving, dyeing and finishing, cotton yarn, and cotton cloth are done in "Greater Manchester"; chemical manufactures are important at Clayton and Blackley especially; there are paper-mills, rubber works, clothing factories, flour-mills, biscuit works, soap works, leather works, and many minor industries. Manchester is now a vast engineering centre, and has been claimed to be the greatest centre in the world for varied engineering.

Manchester (Mancenion) was probably an important settlement in pre-Roman times. The names Irwell and Irk are Celtic. The Romans had a fort at Mancunium, and made it the most important point on their great military road from Chester to York. The Saxon name for the settlement was Mancestre. The most important part of this settlement was at the junction of the Irk with the Irwell, that is close to where the Cathedral, the Grammar School, and Chetham's Hospital and Library now stand. Manchester is men-tioned in Domesday Book and a brief account of its revenues is given. In 1222 there is mention of an annual fair, and Henry III confirmed its charter in 1227. At the end of the thirteenth century cloth-weaving had already been established; and in the next century this industry received a great impetus when Edward III settled some of the Flemish weavers in the town. In the later Middle Ages a " Barons' Hall " stood probably where Chetham's Hospital now stands. Close by was the parish church of St Mary (later to become the Cathedral). The town lay to the east and south and was reached by a bridge over Hanging Ditch (the name of which is still used). In 1515 Hugh Oldham, Bishop of Exeter, founded the famous Manchester Grammar School. Leland visited the town in 1535, and he says "Mancestre, on the south side of the Irwell river, stondeth in Salfordshiret, and is fairest, best builded, quickest, and most populus towne of al Lancastreshire." Camden in 1590 mentions that Manchester woollen cloths are called "Manchester Cottons." It was about the middle of the seventeenth century that the cotton trade began to grow rapidly, cotton-wool (raw cotton) being then increasingly imported from Asia Minor and the Levant. The first Exchange dates from 1729, when the population was 15,000. The Mersey and Irwell had been improved for navigation in 1721; the Bridgewater Canal was commenced in 1760, opened to Manchester in 1761, and completed in 1772; the first modern mill was erected in 1781; the inventions of Kay, Hargreaves, Arkwright, Crompton and Cartwright in connection with spinning and weaving, the work of James Watt on the steam engine, all assisted the rapid growth of the industry and of the town. In 1801 the population had grown to 70,000 and in 1821 it had more than doubled again. The first great railway was finished in 1830, and the furious rate of "progress" was maintained. The increasing specialisation of the cotton industry gradually made of Manchester more and more a market, and distributing centre. The completion of the Ship Canal in 1894 made Manchester into a great port, and to-day it is third or fourth in the kingdom.

Among the educational institutions of the city, the Grammar School has already been mentioned; it has now nearly 1200 boys and about 60 masters. The Manchester University has risen on Owen's College as a foundation. From 1880 to 1903 Manchester was the centre of a Federal University which eventually included Liverpool and Leeds. In 1903 and 1904 Liverpool and Leeds were constituted separate Universities and Manchester continued as "The Victoria University of Manchester." The fine Municipal School of Technology is now a constituent college of the University, and a Faculty of Technology grants degrees in that branch of learning.

The city has many libraries. There is a General Reference Library in Piccadilly, and a Public Commercial Reference Library in the Royal Exchange. The John Rylands Library is housed in a splendid building somewhat hidden by the neighbouring buildings in Deansgate. In this library is a very fine collection of treasures of literature, among which are the Althorp Library (including several Caxtons), the Crawford manuscripts, over 3000 volumes

The John Rylands Library

printed before 1500, many original editions of famous works, and a collection of manuscript and printed copies of the Bible ranging from the sixth century. Chetham's Hospital and Library were founded by Humphrey Chetham, a wealthy fustian manufacturer and merchant. He left an endowment for a hospital for the support and education of 40 poor boys, and for the establishment of a *free* library. The latter is probably the oldest free library in Britain, and one of the oldest in Europe. The collection of manuscripts and books on local history, antiquities, and local topography is very fine. The buildings have been little altered from the seventeenth century, the cloisters being perhaps the most characteristic part.

Other famous buildings and institutions are: the Art Gallery in Moseley Street, the famous Free Trade Hall in Peter Street, the Assize Courts in Strangeways, and the Athenaeum in Princess Street. The Town Hall in Albert Square is a fine building designed by Waterhouse. The Cathedral is the old parish church, which received a collegiate charter in 1421. Manchester was made a Bishopric in 1847. (pp. 1, 5, 7, 16, 22, 26, 27, 34, 39, 54–59, 63, 67, 69, 73, 75–81, 85, 86, 88, 89, 93, 95, 97–106, 108–111, 115–120, 124, 125.)

Middleton, N.C.B. (28,309), lies 5 miles north-north-west of Manchester. Its name is appropriate to its position as it occupies a middle position to Manchester, Oldham, Rochdale and Bury. It has cotton-spinning, cotton-weaving, silk-weaving, calico-printing, bleaching and dyeing, soap-making, jam manufacture and other minor industries. The church is one of the most interesting in the county, and has many associations with the famous Asshetons of Middleton. (pp. 26, 27, 63, 76, 77, 85, 94–96, 105.)

Moseley, N.C.B. (12,705), lies in a deep valley close to the Yorkshire and Cheshire boundaries, 4 miles south-east of Oldham. The industry is chiefly cotton-spinning. (pp. 85, 111.)

Nelson, N.C.B. (39,839), lies 4 miles north-north-east of Burnley. It lies at and near the junction of Colne-water, Pendle-water, and Walverden Brook—all feeders of the Lancashire Calder. Its industry is almost entirely the weaving of high-grade cotton goods. It is of very modern growth and the name only became recognised in the middle of the last century when the railway to Colne was completed. The station was named after the Lord Nelson inn, a wayside inn which had been built at the cross-roads and round which the little hamlet was already growing. Within the borough is the hill known locally as Tum Hill (possibly a corruption of the Latin *tumulus*), another part of the same hill is called Caster Cliff (see Colne); this is also within the Nelson boundary. (pp. 12, 13, 16, 20, 24, 25, 38, 51, 59, 71, 82, 83, 85, 93, 105.)

Newton-le-Willows (or Newton-in-Makerfield), U.D. (18,776), is almost midway between Manchester and Liverpool,

The Infirmary, Oldham

about 16 miles from each. The industries are coal-mining, iron-working, cotton-bleaching and calico-printing, and glass manufacture. Newton was the head of a "Hundred" at the time of the Domesday survey. (pp. 77, 101.)

Oldham, C.B. (145,001). This important town lies about 6½ miles north-east of Manchester on the western slopes of the Pennines. Its chief industry is cotton-spinning, in which it stands first in the world. It uses American cotton chiefly, but the spinning of Egyptian has made great strides during recent years. There are over 17,000,000 spindles in Greater Oldham, out of 40,000,000 in the Manchester province. Other industries are silk-weaving, the making of velvets and fustians, and the weaving of shirtings and sheetings. There is also a little woollen manufacture. Very large engineering works are situated in the town, one of which is said to employ 12,000 persons. The chief production is textile machinery. There is also considerable coal-mining. Oldham is almost entirely of modern growth, its population in 1801 being only 12,024. (pp. 5, 7, 16, 19, 26, 27, 38, 51, 58, 71, 72, 77, 79, 81, 85, 86, 93, 95, 97, 105, 110, 111, 118.)

Ormskirk, U.D. (7407), lies on the West Lancashire plain 12 miles north of Liverpool, 16 miles south-west of Preston. It is surrounded by a wide agricultural district and is in consequence an important market town; it is especially well known as a market for potatoes. The church has many features of interest. It is unusual in having both a square tower and a spire. Burscough Priory (in ruins) is about 2 miles away. (pp. 7, 16, 22, 55, 70, 90, 91, 95, 105, 110.)

Oswaldtwistle, U.D. (15,107), is contiguous with Accrington and Church, the whole forming practically one town. The industries are cotton-spinning, cotton-weaving, calico-printing, chemical manufactures, and brick and tile making. (p. 24.)

Padiham, U.D. (12,474), is 3 miles west of Burnley, on the Calder. Its industries are coal-mining, cotton-weaving, and engineering. There are two famous houses in the district, Gawthorpe Hall and Huntroyde. The former was built at the beginning of the sixteenth century, and was thoroughly restored in 1850. (pp. 24, 25, 85, 91, 103.)

Prescot, U.D. (9043), is about 3 miles south-west of St Helens and 7 miles from Liverpool. It is an old market town with modern industries. Watch-making was formerly important; it then declined, but is now reviving again. Electric cables are made here. (pp. 80, 90.)

Prestwich, U.D. (18,750), is practically a suburb of Manchester. Heaton Park, the property of the Corporation of Manchester, is about a mile to the east.

Radcliffe, U.D. (24,677), is about 7 miles north-west of Manchester, and near the junction of the Roche with the Irwell. Its industries are coal-mining, generating electric power, cotton-weaving, bleaching and dyeing, chemical manufacture, paper-making and engineering. Radcliffe Tower, a ruin now used as outbuildings for a farm, was formerly the home of the Earls of Sussex and Derwentwater. (pp. 26, 27, 76, 85.)

Ramsbottom, U.D. (15,370), is 3½ miles north of Bury, on the banks of the Irwell. It has cotton-weaving, spinning, calico-printing, bleaching, dyeing, and paper manufactures. On the hillsides are two towers of note; that on the east is Grant's Tower, and celebrates the brothers Grant, who were successful manufacturers and merchants, and the originals of the Cheeryble Brothers in Dickens's *Nicholas Nickleby*. The tower on the west is called Peel Tower, in memory of the great statesman. (pp. 21, 26, 27, 76, 85.)

Rawtenstall, N.C.B. (28,381), is 4 miles north of Ramsbottom and in the heart of the Rossendale Fells. The houses and factories are closely packed along the narrow valleys, with broad moors rising to over 1000 feet, lonely and bleak. The industries are cotton-spinning, cotton-weaving, calico-printing, bleaching and dyeing, paper-making, the making of felts, and slipper-making. (pp. 16, 26, 27, 85, 88.)

Rishton, U.D. (7016), is about 2 miles east-north-east of Blackburn on the Leeds and Liverpool Canal. It has cotton manufactures, calico-printing, and paper-mills. (pp. 24, 44, 103.)

Rochdale, C.B. (90,807), is 10 miles northward of Manchester, at the southern edge of the Rossendale Fells. It is drained by the Roche and its tributaries. It has a considerable woollen manufacture, the goods made including blankets, flannels for shirtings, etc., cloth for ladies' dresses, grey flannels for the army, and white flannels for the navy. The cotton trade has however long since passed the woollen trade and now gives employment to seven times as many people. Cotton-spinning, especially fine counts, and doubling; dyeing and finishing cotton goods, engineering, and the making of asbestos and rubber goods are also carried on. Rochdale is a town of some fame because of its connection with great men and great movements. John Bright was born there of a manufacturing family in 1811. The Rochdale Equitable Pioneers started the Co-operative movement in 1844. Among modern buildings of note are the Town Hall, the Art Gallery, and the Free Library. The parish church of St Chad dates from the twelfth century, and contains many portions of ancient structure. (pp. 5, 7, 14, 16, 19, 21, 26, 27, 38, 51, 58, 71, 72, 76-82, 85, 86, 88, 91, 93, 94, 97, 101, 102, 105, 110, 119, 122.)

Royton, U.D. (17,207), is just north of Oldham, and has large spinning-mills. (pp. 26, 59, 71.)

St Helens, C.B. (102,675), is 10 miles east-north-east of Liverpool. The Sankey Canal connects it with Warrington and Widnes. The industries are coal-mining and iron-founding, but the glass manufacture is the most important. There are also chemical works and manufactures of earthenware. A grammar school at the neighbouring village of Farnworth was endowed by William Smyth (1450-1514), Bishop of Lincoln, who also founded Brasenose College. The chief modern buildings are the Town Hall and the Gamble Institute. Three miles to the north is Knowsley Hall, the historic seat of the Earl of Derby. (pp. 5, 7, 16, 71, 72, 75, 77, 80, 104, 105, 111.)

Salford, C.B. (234,150), is divided from Manchester in part by the River Irwell. On the whole it occupies the north-west and west of the great twin town Manchester-Salford. Except for purposes of Local Government the two towns form one great city of nearly a million inhabitants. The railways and canals are common to the two towns, but the Ship Canal Docks are in Salford. The industries are those of Manchester. Salford is a "Royal Borough," because the King is Lord of the Manor. It is an old town and gave its name to one of the hundreds of Lancashire. It has been a borough since 1230, and the charter of that date is preserved in the Town Hall. Simon de Montfort was one of the signatories. In Peel Park are the Museum and the Art Galleries, and in Bulle Hill Park the Natural History Museum. The Royal Technical

Lord Street, Southport

Institute in Peel Park gives instruction in Science, Art, Commerce and Technology. (pp. 3, 5, 26, 27, 79, 85, 89, 104–6, 110, 116.)

Skelmersdale, U.D. (6687), lies about 7 miles west of Wigan. Its industry is coal-mining, and there is good agricultural land all around.

Southport, C.B. (76,664), lies on the south side of the Ribble estuary. From a mere hamlet it has grown in a hundred years into a well-built fashionable town, not only largely frequented by visitors, but a residential town of the front rank. Its climate is mild, and the rainfall comparatively low. The sea is not so much in evidence as at many seaside resorts, for the sands are very extensive. To reach it there is a long pier with a tramway service. It is a popular idea in Lancashire that " the sea is leaving Southport." This means of course that the sands of the Ribble and the Mersey have made considerable additions to the coast hereabouts. The authorities have used these sands with great effect in making the marine lake and gardens between the promenade and the sea. Southport is well laid out, and many of its streets have avenues of trees. Lord Street is considered one of the finest streets in Britain, one side being occupied by shops, the other being a boulevard with many public buildings. (pp. 5, 7, 19, 29-31, 35, 37, 50, 58, 70, 97, 110, 111.)

Standish, U.D. (7294), is 3½ miles north of Wigan, a mining village set in quite pretty rural scenery. The church is noted, having been rebuilt in the reign of Elizabeth. (p. 101.)

Stretford, U.D. (46,535), is 4 miles south-west of Manchester, where the road from Manchester to Chester crosses the Mersey. The Bridgewater Canal passes through it. It is practically a suburb of Manchester, and shares in some of the industries of the latter. (pp. 26, 59.)

Swinton and Pendlebury, U.D. (30,924). This dual town lies to the north-west of Manchester. The industries are coal-mining, cotton-spinning, and chemical manufactures.

Tottington, U.D. (6762), lies north-west of Bury, of which it is almost a suburb. It has cotton-mills, paper works, and soap works.

Trawden, U.D. (2762), is 2 miles south-east of Colne under the shadow of the great mass of Boulsworth Hill. It has cotton-weaving and dye works. In Trawden parish, and about 1¼ miles to the east of the village, is Wycoller, a hamlet in a pretty Pennine dale, with the ruins of an Elizabethan house which Charlotte Brontë used as the original of "Ferndean Manor" in *Jane Eyre*. (pp. 24, 39.)

Turton with Edgworth, U.D. (12,157), lies 5 miles to the north of Bolton among the Rossendale Fells. Edgworth has a famous Children's Home and Orphanage. Turton Tower, dating from 1101, has been much altered. It was at one time the residence of Humphrey Chetham, the founder of Chetham's Hospital (see Manchester).

Tyldesley, U.D. (15,651), lies about midway between Manchester and Wigan. The industries are coal-mining, cotton-spinning and engineering.

Upholland, U.D. (5532), lying about 4 miles to the west of Wigan, is probably the least spoilt of the South Lancashire villages. There are coal-mines, coke ovens, fire-brick works and stone quarries. There was a Benedictine Priory here in the fourteenth century which existed until the dissolution of the monasteries. The Priory church became a chapel of the parish of Wigan. (pp. 63, 72, 90, 95.)

Walkden (in the U.D. of Worsley, which see) is 7 miles north-west of Manchester and 4 miles south of Bolton. Here are coal-mines, and extensive cotton-spinning, weaving, and finishing mills.

Walton-le-Dale, U.D. (12,153), 2 miles south-east of Preston, is on the Ribble, and has cotton-spinning and weaving industries.

Wardle, U.D. (4467), adjacent to Rochdale to the north, has cotton-spinning, cotton-weaving, calico-printing, dyeing and finishing, as well as woollen-mills.

Warrington, C.B. (76,811), is almost midway between Liverpool and Manchester. It is on the north bank of the River Mersey at the point where was the lowest bridge until modern times. A great road to the north passed through it, and it has therefore a most interesting history. The industries are engineering, the making of many kinds of wire, chemical manufacture, soap-making, tanning and leather finishing, glass-making, the making of fustians, paper and printing trades, making rubber goods, waterproofing, making cardboard boxes; and it is a great agricultural market. The origin of the borough, the market, and the erection of the bridge date from the thirteenth century. The town figured prominently in the military movements of the seventeenth and eighteenth centuries. It was first held by the Royal forces in the Civil War and was taken by the Parliamentarians in 1643. In 1648 the Scottish forces under Hamilton surrendered there. Cromwell wrote his two historic letters "To the Speaker of the House of Commons" and "To the Honble the Committee at Yorke" from Warrington. In 1745 the bridge was partly destroyed to prevent the crossing of the rebel forces. Warrington has been associated with many famous men and famous movements. Warrington Academy, founded in 1757, was the predecessor of Manchester College, Oxford. John Harrison invented the compensating pendulum there and won the government prize of £20,000 in 1767. John Howard published at Warrington his famous attacks on prison life and administration. The first Total Abstinence Society was founded there about 1830, and the first rate-aided Free Library in England was started there. So much was Warrington associated with learning and with forward movements that, a hundred years ago, it was known as the Athens of England. The town has a massive statue of Cromwell and a fine parish church with a very lofty spire. The handsome Town Hall is an eighteenth-century mansion which was bought by the Corporation. The Boteler Grammar School was founded in 1526. (pp. 3, 5, 26, 28, 58, 59, 63, 66, 67, 75–78, 80, 93, 95, 101, 103, 105, 111.)

Waterloo with Seaforth, U.D. (29,626), is 4 miles north-north-east of Liverpool, of which it is chiefly a residential suburb.

Westhoughton, U.D. (15,593), is midway between Bolton and Wigan. Its industries are coal-mining, cotton-spinning, and calico-printing.

Whitefield (or **Stand**), U.D. (6902), is practically a residential suburb of Manchester and of Bury; from the centre of the former of which it is about 5 miles distant on the north. (See Radcliffe and Prestwich for industries.)

Whitworth, U.D. (8782), is 3 miles north of Rochdale. The industries are those of Rochdale (*q.v.*). There are sandstone quarries in the Rossendale Fells hereabouts. (p. 102.)

Technical College, Wigan

Widnes, N.C.B. (38,879), lies on the north bank of the Mersey where it begins to widen out rapidly, and where are the last bridges—the railway bridge and footbridge, and the Widnes Corporation "Transporter Bridge." The latter spans the Mersey and the Ship Canal. The industries are chemical manufacture, soap-making, oil and paint works, iron works, and copper-smelting. There is a fine Town Hall. (pp. 23, 33, 75–78, 80, 105.)

Wigan, C.B. (89,447), is situated near the edge of the Rossendale Fells, 16 miles west-north-west of Manchester and 16 miles east-north-east of Liverpool. The industries are coal-mining, iron-smelting, engineering, brick and tile making, cotton-spinning and weaving, chemical manufacture, and the production of clothing, including hosiery.

Wigan played an important part in the Civil Wars of the seventeenth century, the Earl of Derby being defeated here in 1651. In 1745 Prince Charlie passed through the town on his triumphal march southwards to Derby and again on his ill-fated retreat to Scotland. Mab's Cross at the top of Standishgate has woven round it a beautiful legend of the Crusades. The parish church is mainly modern, but it includes some ancient elements. There is a Mining and Technical College, and the Free Library is remarkably well equipped. Wigan itself, as might be expected of a mining town, presents some unattractive features, but in the neighbourhood is much pretty scenery. (pp. 5, 7, 25, 58, 66, 70–72, 81, 85, 88, 97, 101–105, 110, 111.)

Withnell, U.D. (3391), is 5 miles north-east of Chorley, at the edge of the Rossendale Fells. The industries are cotton-spinning, weaving, and calico-printing.

Worsley, U.D. (13,929), (includes Walkden, which see), lies about 6 miles west-north-west of Manchester. It was here that Brindley commenced his canal in 1760, which was completed to Manchester in 1761, and to the Mersey in 1772. Lord Francis Ellesmere (Earl of Ellesmere) practically rebuilt Worsley in the years between 1833 and 1857, and he left it somewhat of a model village. (pp. 104, 117.)

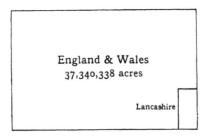

Fig. 1. Area of Lancashire (1,194,555 acres) compared with that of England and Wales

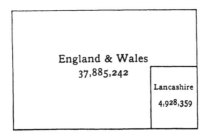

Fig. 2. Population of Lancashire in 1921 compared with that of England and Wales

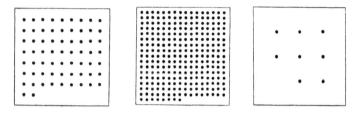

England and Wales 649   Lancashire 2641   Westmorland 83

Fig. 3. Comparative Density of Population to the
square mile in 1921

(*Each dot represents* 10 *persons*)

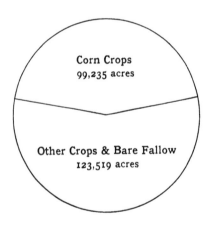

Fig. 4. Proportionate area under Corn Crops compared with
that of other cultivated land in Lancashire in 1926

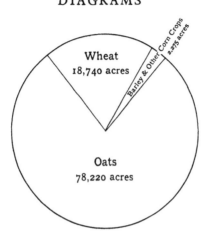

Fig. 5. Proportionate areas of chief Cereals in
Lancashire in 1926

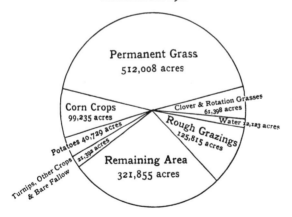

Fig. 6. Proportionate areas of Land in
Lancashire in 1926

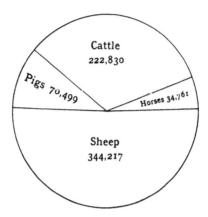

Fig. 7. Proportionate numbers of Live Stock in
Lancashire in 1926

www.ingramcontent.com/pod-product-compliance
Ingram Content Group UK Ltd.
Pitfield, Milton Keynes, MK11 3LW, UK
UKHW042144280225
455719UK00001B/99